Aspects of Microbiology 4

Bacterial Plasmids

Second edition

Kimber Hardy

First published in 1981
Van Nostrand Reinhold (UK) Co. Ltd
Molly Millars Lane, Wokingham, Berkshire, England

Reprinted 1983, 1984
Second edition 1986
Reprinted 1988

Photoset in Times 9 on 10pt by Kelly Typesetting Limited
Bradford-on-Avon, Wiltshire

Printed and bound in Hong Kong

ISBN 0 442 31765 4
(ISSN 0266–6642)

Contents

1 Introduction

Many important bacterial genes are not part of the main chromosome but are on separate circles of DNA called plasmids. A plasmid is a molecule which can be stably inherited without being linked to the chromosome. Some bacteriophages which can also be classified as plasmids according to this definition are considered here only where their study throws light on other plasmids. Plasmids are important in medicine and in agriculture because they confer antibiotic resistance on pathogens of animals and man, and because they can code for toxins and other proteins which increase the virulence of these pathogens. Other plasmid genes are more beneficial. Plasmids enable species of *Rhizobium* to fix nitrogen in the nodules of leguminous plants. They also code for antibiotics which can be used to control pathogenic bacteria. Plasmid genes code for a wide range of metabolic activities and enable bacteria to degrade compounds which would accumulate as pollutants if they were not degraded by micro-organisms. Among the many compounds degraded by plasmid-encoded enzymes are the widely used herbicide 2, 4-D (2, 4-diphenoxyacetic acid) and several components of mineral oils.

Various criteria are used to classify plasmids. The most conspicuous, or the most important, characteristic is used to provide one type of classification. Thus *R plasmids* confer resistance to one or more antibacterial drugs, *Col plasmids* code for anti-bacterial proteins called colicins and *degradative plasmids* code for a variety of catabolic enzymes. *Virulence plasmids* increase the pathogenicity of bacteria in various ways. Plasmids are largely responsible for the virulence of the bacteria causing plague, dysentery, anthrax and tetanus as well as many other diseases of man, animals, fish and plants.

Each of these types of plasmid can be found in a wide range of bacterial genera. Apart from the common feature which is used as the criterion for the classification, many plasmids within each group appear to be totally unrelated to each other. Some belong to several of these groups; some virulence plasmids specify drug-resistance, and many R plasmids also code for colicins.

Bacteria often contain two or more different plasmids which can co-exist, and are said to be *compatible*. *Incompatible* plasmids cannot co-exist together; after a few generations of bacterial growth one or other is lost. Plasmids are classified into incompatibility groups on this basis. Those in the same group are also usually related to each other in other ways.

Plasmids can be either conjugative or nonconjugative. *Conjugative plasmids* transfer copies of themselves from one bacterium to another and many of them are known to code for protein tubes called sex pili. DNA is believed to pass through these tubes from one cell to another. Conjugative plasmids can also transfer pieces of chromosomal DNA between bacteria, and are therefore sometimes called *sex factors*. These plasmids are very useful for mapping the positions of chromosomal genes.

Insertion sequences and *transposons* are genetic elements which frequently occur on plasmids. They also occur on bacterial and phage chromosomes, and range in size from about 1000 to 5700 base pairs. Their peculiar feature is that they are able to transfer themselves or a copy of themselves to DNA molecules with which they have little or no sequence homology. Transposons code for many of the important characteristics specified by plasmids such as antibiotic resistance and toxin formation. Insertion sequences are also responsible for several important properties of plasmids; for example, they enable some plasmids to become part of the bacterial chromosome. Plasmids which can exist either autonomously (separate from the chromosome) or integrated into the chromosome are called episomes.

Plasmids are very useful for studying the properties of bacterial genes and they are also becoming indispensable for studying many aspects of eukaryotic genes. They are used as *vectors* to *clone* DNA. A variety of different enzymes can be used to insert pieces of DNA, from animals, plants or prokaryotes, into plasmids. Circular molecules consisting partly of plasmid DNA and partly of inserted DNA can then be put back into a suitable bacterium. The plasmids replicate during bacterial growth so that many copies of the cloned DNA can easily be obtained. Several eukaryotic proteins have been successfully made by bacterial cells by the transcription and translation of genes cloned in plasmid vectors.

Plasmids found in *Escherichia coli* and related members of the *Enterobacteriaceae* have been the subject of most of the research on the structure and replication of plasmids (Chapter 2), and on their ability to transfer copies of themselves by conjugation (Chapter 3). Plasmids from a wider range of bacteria are described in Chapters 4 and 5, where plasmid genes of medical, agricultural and environmental importance are considered.

The plasmids referred to in this book are listed in the appendix; some of the names can be confusingly similar to someone unfamiliar with the literature. It has not been possible to cite references to individual papers here, so the policy in most cases has been to refer to the most up-to-date and most readily available reviews, where references to original papers can be obtained.

2 Structure and replication

Bacterial plasmids are molecules of double-stranded DNA. No RNA bacterial plasmids have been found, although several fungal plasmids and numerous bacteriophage chromosomes are composed of RNA. All the bacterial plasmids examined so far exist predominantly as circular molecules in their host cells.

The molecular weights of plasmids range from about 1×10^6 to 200×10^6. Plasmids therefore range in size from about 0.04% to 8% of the *E. coli* chromosome, (molecular weight of about 2.7×10^9; length of about 1.3mm). The relationship between molecular weight, length in μm, number of base-pairs and potential coding capacity of DNA is listed in Appendix 1. Plasmids which have molecular weights of more than 100×10^6 are almost exclusively found in Gram-negative bacteria, particularly in strains of *Pseudomonas* and *Agrobacterium*. Two small linear plasmids were recently discovered in *Streptomyces rochei*.

Circular and other forms of plasmid DNA Most of the plasmid DNA inside bacteria is in the form of a *covalently-closed circle* (CCC), meaning that there are no breaks in either of the two polynucleotide strands which comprise the double-helix. Most of the CCC plasmid molecules isolated from bacteria are twisted to form *supercoiled* molecules which have superhelical twists (Fig. 1). For example, a small plasmid with a molecular weight of 2.1×10^6 (3200 base-pairs) which was isolated from *E.coli* had an average of 19 superhelical twists per molecule at 37°C in 0.2M NaC1. Plasmids isolated from bacteria often appear in the electron microscope as tightly coiled and branched structures because of these superhelical twists.

The formation of superhelical twists is not an intrinsic property of CCC plasmids. If the ends of a linear DNA molecule are simply joined together to form a circle, the circle does not become supercoiled unless the two strands at one end are rotated about each other before being joined to the two strands at the other end. If one end is rotated in this way, the circular molecule rotates in the opposite direction to form superhelical twists which relieve the strain. Supercoiled plasmids isolated from bacteria have *negative* superhelical twists. Negative twists are introduced into a circular molecule if one of the polynucleotide strands is broken and the double helix is partially *unwound* before the broken strand is rejoined. Enzymes called *gyrases,* which are involved in plasmid replication, are responsible for introducing negative superhelical twists into plasmid DNA molecules.

If one of the two polynucleotide strands in a closed-circular plasmid is broken, or *nicked,* an *open-circle* is formed. If the CCC molecule was previously supercoiled, the superhelical twists are lost—the molecule unwinds and becomes *relaxed.* When both polynucleotide strands are broken a linear molecule is formed if the two breaks are either exactly opposite, or so close together that the hydrogen bonds between the intervening complementary bases are not strong enough to hold the two strands together. The small proportion of open-circular

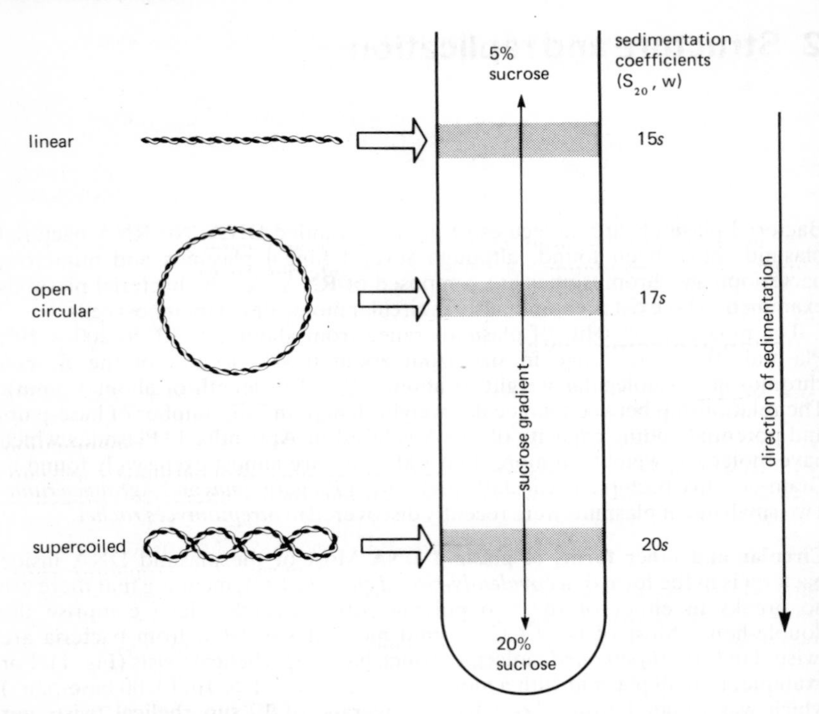

Fig. 1 Circular and linear forms of plasmid ColE1

and of linear molecules found in cell extracts may be derived in part from CCC molecules which were nicked when the cells were broken. Some of the very large plasmids are particularly difficult to keep in the CCC form during isolation and purification.

Other forms of plasmid DNA also occur in cell lysates, including dimers, trimers and other multimers. Catenanes (interlocked rings) also occur. These are usually rare and are believed to result from errors in replication.

Sedimentation coefficients of the various forms of plasmid DNA Supercoiled CCC, open-circular and linear plasmid DNA have different sedimentation coefficients, and can be separated according to their sedimentation velocities in sucrose density gradients. In this procedure, a mixture of the three forms of plasmid DNA is layered onto a sucrose density gradient—for example, a gradient from 5% (w/v) sucrose at the top of the tube to 20% (w/v) sucrose at the bottom of the tube. When the tube is centrifuged, the supercoiled CCC form of the plasmid moves at the greatest velocity, followed by the open-circular form and finally the linear form (Fig. 1). Sedimentation coeffients (S_{20w}) for the three forms of plasmid ColE1 (Molecular weight 4.3×10^6) under these conditions are 23*S* (CCC form), 17*S* (open circular form) and 15*S* (linear form).

Under alkaline conditions which denature DNA molecules (that is, which separate the two polynucleotide strands of a double-helix) differences between the sedimentation coefficients of the CCC and the other forms become more pronounced. When the hydrogen bonds between complementary bases in open-circular or linear molecules are broken at high pH, the two polynucleotide strands of each plasmid can unwind completely to form two separate strands. But the two polynucleotide strands of a CCC molecule cannot separate completely; the molecule collapses to form a highly compact structure, which has a correspondingly high sedimentation coefficient. Neutral and alkaline sucrose gradients can also be used to isolate dimers and catenanes, which have characteristic sedimentation coefficients (see Freifelder, 1976).

Isolation of plasmid DNA

Most of the methods used to isolate plasmids depend on their small size in comparison with the bacterial chromosome, and on their circular form. The major difficulties are to ensure that the plasmid DNA is free from contamination by chromosomal DNA, and to avoid the breakage and loss of plasmids, especially large ones, when the cells are broken open and their contents centrifuged to remove debris. Fortunately, many of the methods which were developed for isolating the circular molecules of viral and mitochondrial chromosomes are applicable to plasmids, so that their isolation is now relatively straightforward. Small plasmids related to ColE1, which are frequently used as vectors in genetic engineering, are particularly easy to isolate because they continue to replicate for several hours in the presence of chloramphenicol (unlike the *E.coli* chromosome). When protein synthesis is inhibited by chloramphenicol, the *E. coli* chromosome completes rounds of replication which may be in progress, but no further rounds begin. ColE1 can initiate many new rounds of replication in the absence of protein synthesis, so most of the DNA in chloramphenicol-treated cultures containing ColEl is CCC plasmid DNA.

There are usually three major steps in isolating a plasmid:

1 Bacteria are broken using a lysis method which does not break CCC plasmid DNA molecules: these methods vary according to the species being investigated. Enterobacteria are usually lysed by first treating them with EDTA (ethylene-diaminetetra-acetic acid) and lysozyme which together break down the cell wall. EDTA disrupts the outer membrane and allows lysozyme to reach and degrade the rigid mucopeptide layer of the cell wall so that sphaeroplasts are formed. In addition, EDTA chelates metal ions and so inactivates cation-dependent nucleases. EDTA also prevents the hydrolysis of phosphodiester bonds. Sufficient sucrose is added to prevent immediate osmotic lysis, so that sphaeroplasts can be lysed in a controlled way by adding a detergent such as sodium lauroyl sarkosinate (SLS).

2 Cell debris and larger fragment of chromosome are removed from the lysate by centrifugation—a 'clearing spin'. Parts of the chromosome remain attached to fragments of cell envelope, and sediment with the debris.

3 CCC plasmid DNA is separated from fragments of chromosome which remain in the supernatant by dye-buoyant density centrifugation. Plasmids

usually have about the same percentage of guanine-cytosine base-pairs as the chromosome of their bacterial host. The two types of DNA molecules therefore have similar buoyant densities and cannot be separated by sedimentation to equilibrium in a density gradient. However, if the plasmid DNA is in the CCC form, substantial differences in buoyant density can be introduced by adding the red dye ethidium bromide. Ethidium bromide binds tightly to DNA by intercalation between adjacent base-pairs, causing the double-helix to unwind; the DNA molecules become longer and their buoyant densities decrease. When added to CCC plasmid molecules, unwinding of the double-helix caused by the dye is accompanied by compensatory superhelical twists in the opposite direction. Ethidium bromide ceases to bind when counter-active effects of the superhelical twists prevent further intercalation of the dye and unwinding of the double-helix. A linear or an open-circular molecule is not subject to the same constraints imposed by superhelical twists; the double-helices of these molecules can unwind further so they bind more ethidium bromide.

At equilibrium in caesium chloride density gradients containing ethidium bromide, CCC plasmid DNA forms the lower of the two bands in the gradient because of its greater buoyant density. (This method does not separate plasmids on the basis of their size or sedimentation coefficients, in contrast to the sucrose gradients described above.) Plasmid DNA is removed from the side of the tube through a needle attached to a syringe, or by collecting sequential fractions from either the top or the bottom of the tube. Ethidium bromide is removed from the plasmid DNA by extraction with an organic solvent, and the DNA is then dialysed into an appropriate buffer. Endonucleases and other proteins which may cause degradation of DNA during storage are removed by extraction with phenol.

The different effect of a high pH (pH12–12.5) on chromosomal and plasmid DNA can also be used to remove most of the chromosomal DNA from lysed cells. High pH denatures chromosomal DNA, but not CCC plasmids. If the pH is subsequently lowered and the salt concentration kept high, chromosomal DNA precipitates, leaving CCC plasmids in solution. Most of the high molecular weight cellular RNA also precipitates under these conditions as well as proteins if the detergent sodium dodecyl sulphate is present. After centrifugation to remove the precipitate, the plasmid DNA is already sufficiently pure (after ethanol precipitation) for some kinds of experiment. It can be further purified by caesium chloride—ethidium bromide centrifugation if necessary.

Methods for determining plasmid molecular weight

Kleinschmidt developed a very useful method for examining DNA molecules by electron microscopy. Plasmid DNA is mixed with a basic protein, usually cytochrome *c*, and a small drop of the mixture is allowed to flow down an inclined microscope slide onto the surface of an appropriate salt solution in a Petri dish (Fig. 2). The cytochrome *c* becomes denatured on the surface and spreads out as a film. It binds to the plasmid molecules which also spread out over the surface. A suitably coated electron microscope grid is touched onto the film so that some of the cytochrome *c* and DNA attaches to it. The grid is then stained with a uranium salt and shadowed with metal (usually a platinum/palladium alloy) to make the cytochrome *c*-coated DNA visible in the electron microscope. The molecular weights of plasmids can be determined from electron micrographs by comparing

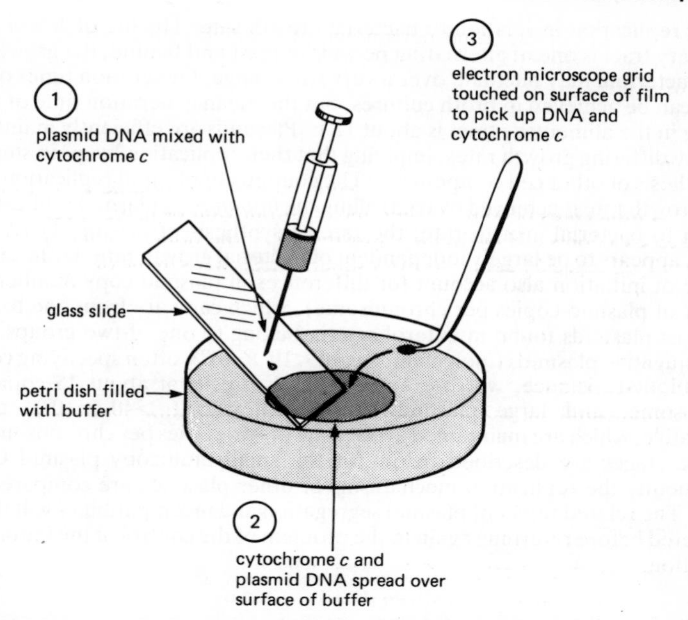

Fig. 2 Kleinschmidt procedure for examining plasmid DNA by electron microscopy

the lengths of open circular molecules with those of other DNA molecules of known molecular weight.

Plasmid molecular weights can also be determined by sedimentation through sucrose gradients or by agarose gel electrophoresis. In both methods, the various forms of plasmid DNA migrate at different rates, so it is important to ensure that the correct form of the plasmid is compared with an appropriate standard which is included in the sucrose gradient or agarose gel. To determine the positions of plasmids after agarose gel electrophoresis, the gels are stained with ethidium bromide. This binds to the DNA and fluoresces when illuminated by ultraviolet light.

Plasmid replication

Many plasmids are maintained so efficiently in bacterial cultures that less than one cell in 100 000 lacks a functional copy of the plasmid. Some, such as RP4, maintain this efficiency in many different genera of Gram-negative bacteria; others are stably maintained in only a few species. *Curing agents* are substances which increase the rate of loss of plasmids during bacterial growth. Acridine orange interferes with replication of the F plasmid and of several other plasmids so that they cannot be maintained efficiently. Another inhibitor of DNA synthesis, mitomycin C, is an effective curing agent for many plasmids found in pseudomonads.

An important aspect of the adaptation of plasmids to their hosts is the control of plasmid replication in relation to bacterial growth rate. The life of *E.coli* in the alimentary tract is one of alternating periods of feast and famine; the growth rate of the bacterium may fluctuate over a very wide range. Generation times of only 20 min can be achieved in broth cultures, but the mean generation time of *E.coli* growing in the alimentary tract is about 12 h. Plasmids are efficiently maintained at widely differing growth rates, implying that their replication keeps in step with the synthesis of other cell components. The coupling of plasmid replication to the host's growth rate is achieved by controlling the *initiation* of plasmid replication in relation to bacterial growth rate; the rate of synthesis of plasmid DNA, once started, appears to be largely independent of bacterial growth rate. Differences in the rate of initiation also account for differences in plasmid copy numbers (the number of plasmid copies per chromosome), which can vary from one to about 200. Most plasmids found in enterobacteria belong to one of two groups: small nonconjugative plasmids (molecular weight $<10 \times 10^6$), often specifying colicins or antibiotic-resistance, which are maintained in cells at about 15 copies per chromosome; and large plasmids (molecular weight $>40 \times 10^6$), usually conjugative, which are maintained at only one or two copies per chromosome.

These stages are described below for the small multicopy plasmid ColE1; subsequently the replication mechanisms of other plasmids are compared with ColE1. The related topics of plasmid segregation and incompatibility will then be considered before returning again to the problem of the control of the initiation of replication.

Replicaton of ColE1

ColE1-K30 is a relatively small plasmid which is maintained in strains of *E.coli* at about 15 copies per chromosome. It has a molecular weight of 4.3×10^6 and could therefore code for eight different proteins each with a molecular weight of 30 000 (assuming that there are no overlapping genes). ColE1 codes for colicin E1, an antibacterial protein, and a protein which confers immunity to colicin E1.

ColE1 encodes a small polypeptide which regulates its replication, but it is apparently not essential under most conditions. Thus ColE1 appears to be entirely dependent on host enzymes for its replication.

Plasmid replication is conveniently divided into three major stages: initiation, elongation of polynucleotide chains by semi-conservative synthesis, and termination.

Initiation of ColE1 replication Replication begins at a specific point in the plasmid, the origin, and then continues in one direction around the circular molecule so that the replication fork (Fig. 3) finally returns to the origin. ColE1 replication is therefore *unidirectional*.

The results of experiments by Tomizawa and his co-workers on ColE1 replication in a cell-free system emphasize the important role played by RNA and the host's RNA polymerases in the initiation of ColE1 replication and in the control of this process. The first event appears to be the transcription of DNA to form an RNA molecule called RNA II (or primer transcript). The ColE1 DNA is transcribed by the host's RNA polymerase beginning at a point 555 nucleotides before the replication origin (that is, upstream from the origin) (Fig. 2).

Most of the RNA transcripts extend past the origin of replication, terminating at a variety of different points. About half of the RNA II molecules hybridize to the complementary template DNA near the origin and are then cleaved at the origin by RNase H (a host-encoded enzyme specific for DNA:RNA hybrids). Cleavage by RNase H exposes an RNA end at the origin and this serves as a primer at which DNA synthesis can begin. Cleavage of the RNA II transcript is precise to within two or three nucleotides, so DNA synthesis almost always begins at one of three nucleotides in ColE1. Presumably, the precise cleavage of the transcript by RNase H depends on the folding of RNA II, i.e. that it has the appropriate secondary structure.

The RNA primer provides a polynucleotide having a 3′ hydroxyl group at one end; an end of this type is essential for initiation of DNA synthesis by all DNA polymerases found in *E.coli.* DNA synthesis which begins at the 3′ OH end of the primer continues for about 500 nucleotides to form part of the new L-strand of ColE1 (Fig. 3). (ColE1 can be separated into *heavy* (H) and *light* (L) strands in caesium chloride density gradients containing poly (U,G)). This reaction is catalysed by DNA polymerase I. DNA gyrase and other proteins assist replication by unwinding the double helix and by binding to single-strand DNA at the replication fork.

Only a small region of ColE1, comprising about 580 base pairs, is essential for replication. This fragment includes the origin of replication and the RNA II promoter. No more than 13 of the bases which come after the origin are essential for replication.

Chain elongation Once the initial part of the new L-strand has been formed, synthesis of the new H strand begins. This strand is synthesized discontinuously in the 5′ to 3′ direction. DNA fragments, called *Okazaki fragments* after their discoverer, of about 1000 bases are formed and are subsequently linked together. Synthesis of each DNA fragment is initiated at a short primer of RNA. The *dnaG* gene-product (DNA primase) transcribes ColE1 to provide this primer. Synthesis of each DNA fragment is initiated at a short primer of RNA. The *dnaG* gene-product (primase) transcribes ColE1 to provide this primer. *E.coli* DNA primase forms part of a complex structure (a 'primosome') composed of about 20 polypeptides, including the *dnaB* and *dnaC* proteins. As the primosome moves along the DNA at the replication fork (using energy from ATP), the *dnaB* protein forms a secondary structure in the single-stranded DNA, enabling the primase to initiate transcription. The *dnaB* gene product seems to be a mobile promoter which enables the primase to initiate transcription. RNA primers are extended by DNA polymerase III holoenzyme (the *dna E*-gene product and at least seven other sub-units, including the products of genes *dnaQ, dnaX, dnaZ* and *dnaN*).

The new L-strand is probably not synthesized discontinuously, i.e. the initial 500 nucleotides is simply extended continuously in the 5′ to 3′ direction as shown in Fig. 3. However, it is known that DNA polymerase III holoenzyme, rather than DNA polymerase I, synthesizes the remainder of the new L-strand once the first 500 nucleotides have been formed.

ColE1 replication also requires DNA gyrase and single-strand binding protein. The two strands of a circular DNA molecule must be unwound to allow replication. This unwinding would cause the unreplicated part of the circle to become supercoiled unless there was a *molecular swivel* to remove the twists.

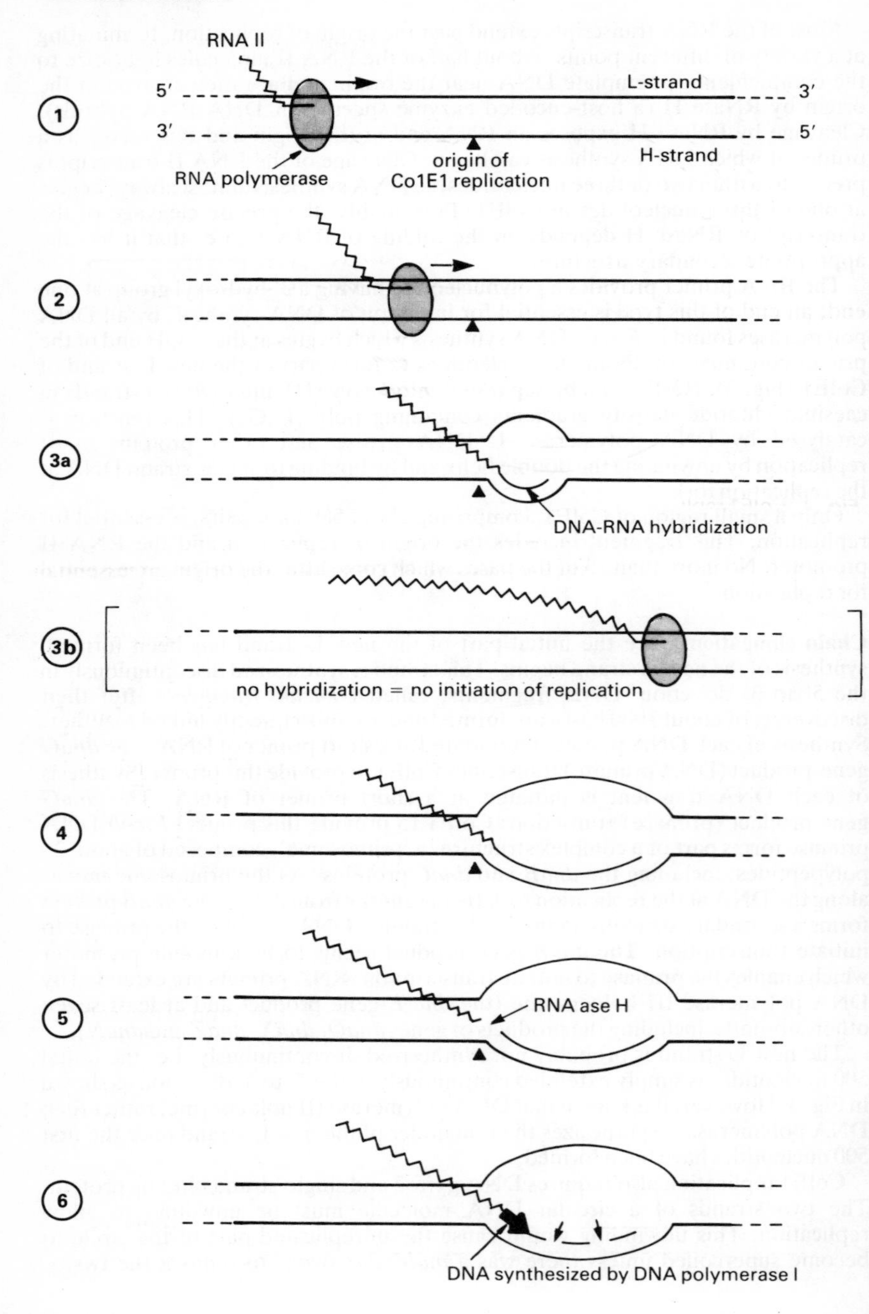
RNA II
5′
L-strand
3′
1
3′
5′
H-strand
RNA polymerase
origin of
Co1E1 replication
2
3a
DNA-RNA hybridization
3b
no hybridization = no initiation of replication
4
5
RNA ase H
6
DNA synthesized by DNA polymerase I

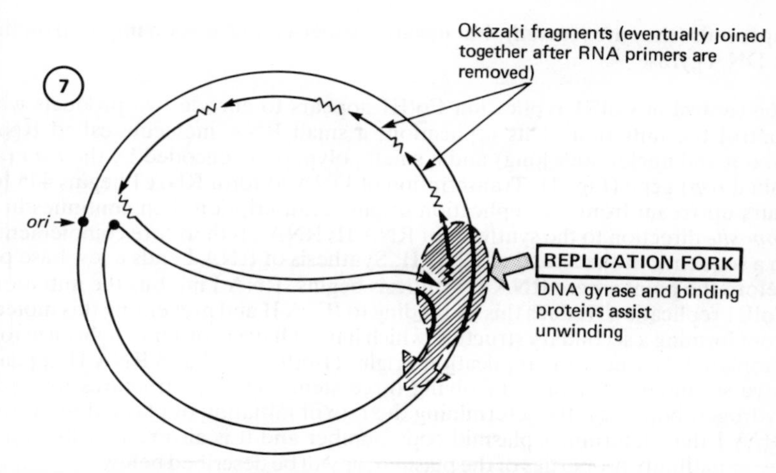

Fig. 3 ColE1 replication. 1. Transcription initiated by RNA polymerase. 2. Elongation of RNA II. 3a. RNA II hybridises with DNA at the origin. 3b. If RNA II does not hybridise at the origin, DNA synthesis is not initiated. 4. Elongation of the DNA:RNA hybrid. 5. Cleavage by RNase H. 6. DNA synthesis begins at the end of the cleaved RNA primer. 7. The replication fork proceeds in one direction around ColE1. (partly redrawn from Masukata, H. and Tomizawa, J.-I. *Cell* 36, 515–522 (1984).

Without such a swivel, replication would stop once the energy of supercoiling counteracted the effect of proteins which unwind the double helix (see also p.3). The molecular swivel appears to be DNA gyrase. Indeed, this enzyme probably actively assists replication by inducing negative supercoiling ahead of the replication fork. (DNA gyrase imposes negative supercoiling on closed circular molecules by introducing nicks in the strands, unwinding the double-helix and then re-joining the strands.)

RNA primers at the 5′ ends of Okazaki fragments are removed and replaced by DNA. The DNA polymerase I protein is probably responsible for both these reactions; it has a 5′ → 3′ exonuclease activity which could remove the RNA primers and a 5′ → 3′ polymerase activity which could replace the RNA by DNA. Adjacent DNA fragments are finally linked together by DNA ligase.

Termination The final stage in Co1E1 replication is the formation of two separate ColE1 molecules, each a covalently closed circle. The two parental strands of ColE1 (that is, those which serve as templates for semi-conservative synthesis) separate before the replication fork reaches the terminus/origin of replication. Separation is probably brought about by the nick-unwind-rejoin activity of DNA gyrase. When the parental double-helix is almost completely unwound and the replication fork has almost reached the terminus, the effect of DNA gyrase activity is to completely separate the two parental strands and hence the two daughter ColE1 molecules. The gap in the newly synthesized strand of each molecule is then closed in a reaction which requires DNA polymerase I and DNA ligase. Immediately after the gap is closed, the molecules are briefly without supercoils (they are *relaxed*), but negative super-coiling is soon imposed on them by DNA gyrase.

The control of ColE1 replication ColE1 appears to encode two products which control the initiation of its replication, a small RNA molecule called RNA I (about 180 nucleotides long) and a small polypeptide encoded by the *rom* (also called *rop*) gene (Fig. 4). Transcription of DNA to form RNA I begins 445 base pairs upstream from the replication origin. Transcription then continues in the *opposite* direction to the synthesis of RNA II; RNA I is therefore complementary to a sequence of the 5′ end of RNA II. Synthesis of RNA I ends a few base pairs before the point where RNA II synthesis begins. RNA I inhibits the initiation of ColE1 replication. It does this by binding to RNA II and preventing this molecule from forming a secondary structure which it must have in order to hybridize to the template DNA near the replication origin. (Both RNA I and RNA II appear to have secondary structures involving three stem-and-loop structures formed by hydrogen-bonding). By determining the rate of initiation of plasmid replication, RNA I thus determines plasmid copy number and it is also responsible for the incompatibility properties of the plasmid, as will be described below.

A region *downstream* from the ColE1 origin is also involved in the control of the initiation of plasmid replication and in the regulation of plasmid copy number. Derivatives of ColE1 which had a deletion of a region downstream from the origin had a higher copy number, so it was proposed that this DNA segment might encode another inhibitor of ColE1 replication. The product, a 63 amino acid polypeptide, was subsequently identified as the product of the *rom* (or *rop)* gene. This protein enhances the binding of RNA I to RNA II so that RNA II functions

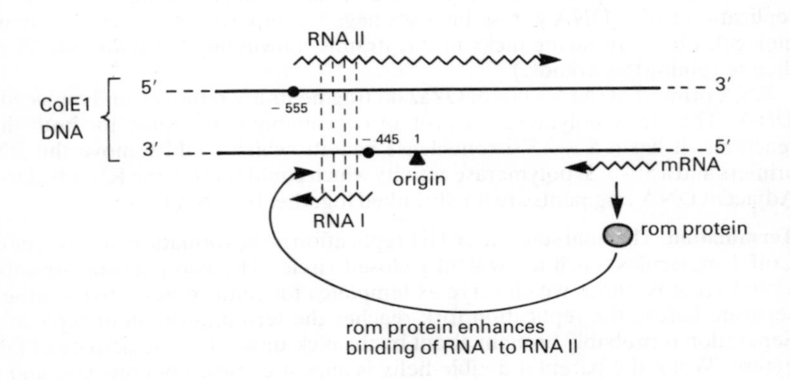

Fig. 4 Control of ColE1 replicaton. RNA II hybridises to the origin and is cleaved by RNAse H to form a primer for DNA synthesis. RNA I binds to RNA II and prevents it binding to DNA at the origin. The *rom* (or *rop*) protein enhances the binding of RNA I to RNA II.

less effectively as a primer. Because it enhances the binding of RNA I to RNA II, the *rom* protein reduces the copy number of ColE1.

Methods used to study ColE1 replication Replication of ColE1 has been studied using intact cells and cell-free extracts of *E.coli*. The approximate position of the ColE1 replication origin and the unidirectional mode of its replication were determined from analyses of partially replicated molecules by electron microscopy. Tracings of molecules in different stages of replication were aligned with respect to the single site in ColE1 which is cut by the enzyme *Eco*R1 (for details of restriction endonucleases, see Chapter 6). Examination of linear molecules produced by *Eco*R1 showed that the replication fork travelled from a unique origin, and always in the same direction. This mode of replication was seen in molecules isolated from cells and from cell-free extracts capable of replicating ColE1. Electron microscopy also showed that in almost all replicating molecules the parental strands remained joined during replication so that θ-shaped structures were formed (see Fig. 3 diagram 7).

ColE1 can replicate in cell extracts which are unable to synthesize proteins and which are prepared from plasmid-free bacteria, indicating that none of the proteins coded for by this plasmid is essential for its replication. A similar conclusion can be drawn from the results of experiments with *phasmids*. Phasmids have been made by linking ColE1 to a phage chromosome. These hybrid molecules can be packaged into phage coats to form particles which can infect cells. A phasmid comprising ColE1 can replicate in chloramphenicol-treated cells although the phage part of the phasmid is incapable of self-replication.

The host enzymes required for each stage of ColE1 replication were determined from studies of replication in mutant bacteria, or in extracts made from such bacteria, and from the effects of enzyme inhibitors on ColE1 replication both *in vivo* and *in vitro*. ColE1 continues to replicate for several hours in cells which have been treated with chloramphenicol to inhibit protein synthesis, but it ceases rapidly when rifampin, an inhibitor of RNA polymerase, is added. Rifampin inhibits synthesis of the first 500 bases of L-strand DNA in cell-free extracts, but not the later stages of ColE1 replication, indicating that a different enzyme catalyses the synthesis of the other RNA primers. The roles of several other proteins were determined from the effects on *in vitro* replicaton of specific enzyme inhibitors or of antibodies against individual proteins. These methods were used to establish the roles of DNA polymerases I and III, DNA gyrase, DNA ligase, DNA primase and the *dnaB* protein.

Most of the bacterial mutants used to study plasmid replication are temperature-sensitive mutants which are defective only at high temperature, i.e. the *restrictive temperature* (usually from 40° to 44°C). The mutants are usually defective in replication of the bacterial chromosome at the restrictive temperature. Genes required for replication of the *E. coli* chromosome are classified into two groups: those which are needed for initiation of replication (such as *dnaA* and *dnaC*) and those which are required for subsequent semi-conservative synthesis of DNA (such as *dnaB, dnaC, dnaE* and *dnaZ*). ColE1 replication is inhibited at the restrictive temperature in *dnaB, dnaC, dnaG, dnaE* and *dnaZ* mutants, indicating that the products of all these genes are required. In addition, ColE1 cannot replicate in *polA* mutants which lack the 5′→3′ polymerase activity of DNA polymerase I, but which retain the 5′→3′

exonuclease activity associated with this enzyme. Replication of the *E. coli* chromosome can occur in *polA* mutants of this type, although Okazaki fragments are joined more slowly than in *polA*$^+$ strains. Cell-free extracts obtained from mutant bacteria can also be used to determine the roles of various enzymes at each stage of replication. Additional information can be gained by adding purified enzymes to extracts of mutant bacteria.

Replication of conjugative plasmids

Most conjugative plasmids found in enterobacteria differ from ColE1 in being maintained at only one or two copies per chromosome, and in being able to replicate in *polA* strains which have only about 1% of wild type levels of DNA polymerase I activity. Furthermore, they specify at least one protein which is essential for their replication. Not all large conjugative plasmids replicate similarly, however. Some have a bidirectional mode of replication and others replicate unidirectionally. The R plasmid R6K differs from other conjugative plasmids in having a high copy number, and in having a complicated bidirectional mode of replication.

Studies of the replication of a large plasmid can be simplified by isolating a fragment containing the sequences essential for replication and forming a small plasmid from it. In practice, this is achieved using restriction endonucleases and DNA ligase. Large plasmids are cut into pieces by a restriction endonuclease. Each piece is then joined to a fragment of DNA which codes for antibiotic-resistance and which cannot replicate itself. The circular molecules formed *in vitro* are then put into a suitable strain of *E.coli* by transformation. Transformants containing a plasmid fragment which can replicate can be selected because the antibiotic-resistance gene attached to the fragment confers resistance on the host strain. The replication properties and copy numbers of the small plasmids formed in this way appear to be similar to the large plasmids from which they are derived. However, this is not always the case; the replication of small derivatives may not always provide a good indication of how the larger plasmid replicates.

Origins and direction of replication Many conjugative plasmids apparently differ from ColE1 in having two potential origins of replication. For example, a second replication origin on the F plasmid can be detected when the primary origin is inactivated.

Replication of the R plasmids R1, R100 and R6-5 resembles ColE1 replication in that it is unidirectional. (These three related R plasmids belong to the incompatibility group FII and their genes for conjugation are similar to those of the F plasmid). Replication of the F plasmid (or more precisely, a small fragment of the F plasmid containing the origin of replication) is predominantly bidirectional; when the usual origin of F replication is deleted, replication proceeds unidirectionally from another origin. (Some deletion mutants of ColE1 can also replicate bidirectionally.) The circular chromosomes of *E.coli, Salmonella typhimurium and Bacillus subtilis* are also replicated bidirectionally. Bidirectional replication implies that the two replication forks meet, presumably at about 180° from the origin. This raises the question of whether there is a specific termination sequence at which the two replication forks meet.

The essential replication fragment of F can replicate efficiently although it does not have the sequence at which termination normally occurs in the F plasmid, indicating that if there is a specific termination sequence, it is not essential. A termination sequence is present in R6K but it does not appear to be essential and it seems to delay, rather than completely stop replication forks. R6K has a number of unusual features. Unlike other conjugative plasmids it has a high copy number (about 20). It has three origins of replication which occur within a 3900 base sequence; *in vivo* one of these origins is used far more frequently than the other. The termination site for R6K replication is not situated at 180° from the origin. Replication is bidirectional and sequential; a replication fork travels from the origin to the termination site, whereupon another replication fork proceeds from the same origin, but in the other direction, to reach the termination site from the other side of the plasmid circle. R6K has so many unusual features that it has been suggested that it arose from the fusion of parts of unrelated plasmids and now has some replication properties of both.

Control of replication of conjugative plasmids Conjugative enterobacterial plasmids depend very largely on host-specified proteins for their replication but, in contrast to ColE1, they code for at least one protein essential for their replication.

The replication of several closely related conjugative plasmids, (R1, R100, R6–5), belonging to incompatibility group FII has been studied in detail, particularly with a view to understanding how their replication is controlled. It appears that once replication has started, subsequent elongation of the DNA chains is catalysed by the same host enzymes which catalyse the replication of the *E.coli* chromosome or plasmids like ColE1.

Replication of Inc FII group plasmid R1 (or NR1) is controlled by regulating the synthesis of the *rep*A1 protein, which is necessary for the initiation of replication at the origin. Synthesis of the *rep*A1 protein is controlled by a small RNA molecule (called either RNA I or RNA E by different research groups). This is the product of the *cop*A or *inc* gene which is transcribed from the opposite DNA strand to that which encodes *rep*A1. (see Fig. 5). RNA I inhibits synthesis of *rep*A1 protein because it binds (by complementary base pairing) to the *rep*A1 mRNA. This alters the folding of the mRNA so that the ribosome binding sites are less exposed and thus translation of the mRNA into the polypeptide *rep*A1 is inhibited. RNA I is produced constitutively in the cell and about 20 molecules are produced for each RNA molecule encoding the *rep*A1 polypeptide. It appears that the regulation of initiation is controlled primarily through the balance of the inhibitor (RNA I) and its target (the *rep*A1 mRNA). As the inhibitor is synthesized constitutively, the amount of it in the cell will be proportional to the number of *cop*A (or *inc*) genes in the cell. Being an RNA molecule, the inhibitor (RNA I) will presumably be unstable and will gradually be degraded by RNAses in the cell so that a steady state concentration should be reached depending on the rates of synthesis and degradation. When a plasmid replicates, the concentration of RNA I should increase (because there are more copies of the gene for RNA I). This should prevent further initiation of replication until the cells grow sufficiently to dilute the concentration of RNA I. The function of *rep*A1 protein in the initiation of replication is unknown.

In addition to this mechanism for controlling replication, the R1 plasmid has a second mechanism. It has been suggested that this might be necessary to ensure plasmid stability because of its low copy number and in particular to act as a rescue mechanism if the copy number should fall too low. The components of this additional mechanism are also shown in Fig. 5. In addition to specifying the *rep*A1 protein, RNA-CX also encodes the *copB* (or *rep*A2) protein which represses the synthesis RNA-II. At the normal copy number, it appears that the concentration of *copB* is sufficient to inhibit transcription of RNA-II so that all the *rep*A1 protein in the cell comes from translation of RNA-CX. But when the copy number is reduced below the usual two per chromosome (experimentally this can be done by putting the genes into the *E.coli* chromosome and monitoring their expression), less *copB* protein is made, so more RNA-II is produced. Since RNA-II encodes the *rep*A1 protein this should stimulate the initiation of replication to bring the copy number up to the normal level.

The control system of R1 thus resembles that of ColE1 in that an essential element in each case is the inhibition brought about by a small RNA molecule when it binds to a homologous RNA transcribed from the opposite DNA strand. A small plasmid from *Staphylococcus aureus*, pT181, also has a control mechanism in which homologous RNA molecules bind to inhibit the formation of a protein essential for replication. Other plasmids do not have replication control mechanisms involving the binding of RNA molecules in this way. These include F, R6K and RK2. In these cases, essential proteins involved in replication have been identified. The F plasmid, for example, appears to encode an essential replication protein which binds near the origin to stimulate replication and also to control its own synthesis. This protein also binds to an adjacent region which may act as a sink to titrate out the protein until overflow molecules become available to act again at the origin.

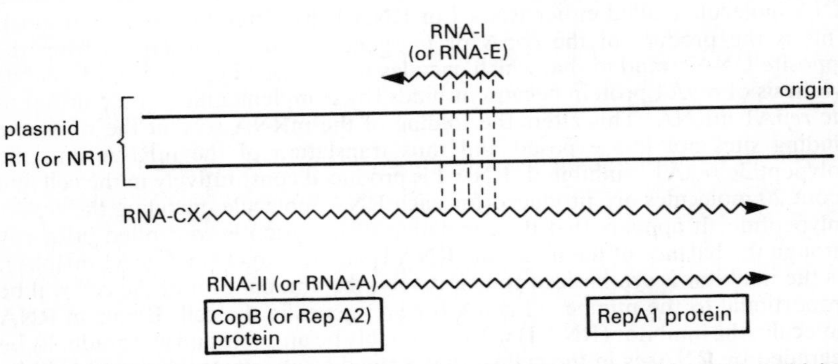

Fig. 5 Control of plasmid R1 replication. Synthesis of the *rep* A1 protein, which is needed for initiating replication, is controlled by RNA-I (also called RNA-E) which binds to RNA-CX and inhibits translation into *rep*A1 protein. RepA2 protein inhibits the formation of RNA II. When the copy number falls below the usual two per chromosome, less *rep*A2 protein is made, so more RNA-II and hence more *rep*A1 protein, is synthesized. Redrawn from Nordström, K., Molin, S. & Light, J. (1984) Plasmid 12:71–90.

Segregation of plasmids at cell division: stability

The stable inheritance of a bacterial plasmid implies that there is an efficient mechanism to ensure that each daughter cell receives at least one copy of the plasmid at cell division. Many plasmid-less cells would soon arise during bacterial growth if low copy number plasmids were simply distributed at random at cell division. The need for an efficient segregation mechanism is presumably less critical for multicopy plasmids. There is no evidence to suggest that ColE1, a multicopy plasmid, has a special partition mechanism.

The segregation mechanism of low copy number plasmids is not understood. The most likely possibility is that plasmids are attached to sites on the cytoplasmic membrane which move into daughter cells at cell division, taking the plasmids with them. One of the features of the model for replicon control proposed by Jacob *et al* (1963) is the important role assigned to the membrane in the control of plasmid replication and segregation.

Genes which increase stability, apparently because they ensure that plasmids are partitioned efficiently at cell division, have been found in several plasmids, including R1, pSC101, prophage P1 and F. These genes are analogous to the centromeres of eukaryotic chromosomes. The *par* genes of different plasmids appear to be interchangeable. For example, the stability of a plasmid R1 replicon lacking its own *par* gene is increased when the *par* locus from pSC101 is added to it. The *par* region of pSC101 does not appear to encode a protein and it is not known how it increases stability. The *par* genes of prophage P1 and F encode proteins. At least three regions of the F plasmid appear to be involved in stability. Two of these genes encode proteins, one of which attaches to a third region of F plasmid DNA which is also involved in plasmid incompatibility. Prophage P1 also encodes a protein involved in stability which specifically binds to a region of P1 DNA. It has been suggested that the *par* proteins which specifically interact with DNA might also bind to sites in the membrane such that they are separated, along with their attached plasmids, into daughter cells at cell division.

In addition to *par* loci, plasmids seem to have at least two other mechanisms to make sure they are not frequently lost from host cells. A mechanism which prevents the accumulation of plasmid multimers has been identified in plasmids P1, a prophage, and ColE1. Prophage P1 exists as a single copy per chromosome. The formation of plasmid dimers (i.e. plasmid circles twice as long as the monomer) by the generalized recombination system of the host, which recombines any homologous sequences, would presumably lead to plasmid loss unless they were resolved again into monomers before the cells divided. A P1-encoded mechanism for rapidly resolving dimers to monomers has been identified. Similar systems are encoded by the multicopy plasmid ColE1 and by F to resolve multimeric plasmids to monomers. The plasmid-encoded systems are independent of the host's *recA* gene-product.

The F plasmid appears to have evolved a particularly ingenious way of increasing its stability. It temporarily inhibits cell division if it has not itself replicated. It seems that if there is only one copy of F in a cell when cell division would normally occur, the plasmid inhibits this division until it has replicated, but it inhibits neither cell growth nor chromosome replication. Thus plasmidless daughter cells cannot be formed. Another F plasmid gene releases the inhibition of cell division once the F plasmid has replicated.

Incompatibility

Bacteria often contain two or more different plasmids which are stably inherited: they are said to be *compatible* with each other. However, certain pairs of plasmids are *incompatible*; if both are added to a suitable host bacterium, by conjugation or some other means, they are unable to co-exist. After only a few generations most of the daughter cells contain only one of the two types of plasmid.

Incompatibility is always seen when genetically distinguishable pairs of the same plasmid are used. For example, two F′ (F-prime) plasmids are incompatible. F′ plasmids are plasmids which have genes normally found on the *E. coli* chromosome attached to them (see p.42 for more details). Examples are F′*his*$^+$, which has genes for histidine biosynthesis, and F′*lac*$^+$, which has genes for lactose utilization. Both plasmids can be transferred into *E.coli,* and a population of cells containing them can be obtained by transferring both F′ plasmids to a strain which is *lac*$^-$ and *his*$^+$. Both plasmids must be present to enable the strain to grow in a minimal salts medium lacking histidine and containing lactose as sole carbon source. However, when no selection pressure is applied (if the strain is grown in nutrient broth, for example) almost all the progeny produced after only a few (say, ten) generations will contain either F′*his*$^+$ or F′*lac*$^+$, but not both.

Incompatibility appears to be a consequence of the mechanisms controlling plasmid replication and segregation at cell division

For plasmids of the ColE1 type, the evidence strongly suggests that incompatibility is caused by RNA I. This molecule interacts with RNA II so that it cannot bind to the DNA at the origin and thus cannot initiate DNA replication. Thus plasmid **A** specifying a RNA II molecule sensitive to the RNA I encoded by plasmid **B** will be incompatible with plasmid B. If plasmids **A** and **B** produce similar RNA I and RNA II molecules, the concentration of RNA I in the cell will maintain a mixture of the two plasmids at the same concentration as if they were alone i.e. if **A** and **B** each have a copy number of 20, the cell will maintain a mixture of **A** and **B** at a total copy number of 20. Eventually, through random sorting of the mixtures at cell division, cells will arise which lack one or other of the two types.

The properties of numerous mutants of F and of IncFII plasmids which have altered incompatibility properties suggest that the mechanisms which control plasmid copy number are also responsible for at least part of the incompatibility of such plasmids.

Plasmids of the IncFII group produce an inhibitor (RNA I) which prevents synthesis of the *rep*A1 protein (see Fig. 5.). A plasmid producing an RNA molecule for *rep*A1 protein which is sensitive to the RNA I encoded by another plasmid should therefore be incompatible with that plasmid. Analyses of mutants of the F plasmid suggest that, in addition to genes which are involved in controlling copy numbers and replication, the genes ensuring that plasmids segregate at cell division also specify incompatibility. Incompatible plasmids might compete for cell components essential for partitioning daughter plasmids at cell division.

Incompatibility is used to classify plasmids into groups; any two members of the same group cannot stably co-exist in the absence of selection pressure. Plasmids found in enterobacteria have been classified into about 25 incompatibility groups.

There are seven incompatibility groups of *Staphylococcus* plasmids and at least eleven groups of *Pseudomonas* plasmids. As many as seven compatible plasmids can be maintained in *E.coli* in the absence of selection pressure; the plasmids in such a strain can account for about 25% of the total cell DNA.

Classifying plasmids according to their incompatibility reactions has proved useful. But difficulties sometimes occur, often because plasmids have two basic replicons and origins of replication and thus express a variety of incompatibility determinants.

Control of plasmid replication

The most influential early model for plasmid replication and for the regulation of plasmid stability and for the regulation of plasmid stability was that proposed by Jacob *et al.* (1963). They defined replicons as 'units of DNA capable of independent replication which set up specific systems of signals allowing, or preventing, their own replication.' Bacterial chromosomes, bacteriophage chromosomes and plasmids are all replicons. They are all capable of autonomous and controlled replication in an appropriate bacterial host cell, in contrast, for example, to a random fragment of chromosomal DNA which is introduced into a bacterial cell by transduction or by some other means. Such a fragment remains unreplicated until it is eventually degraded or becomes part of a replicon by recombination.

Jacob *et al* proposed that the properties of replicons could be explained if each replicon had two specific genetic determinants: a structural gene coding for an *initiator,* which acts at another site, the *replicator,* to initiate replication at the origin (which might be close to the replicator). In this model, the effect of the initiator is therefore essentially positive. However, the experiments which originally suggested a positive control system (rather than a negative system involving a repressor) now seem less convincing in the light of more recent research. The control mechanisms of some plasmids seem to involve both positive and negative elements, so that it is not clear whether the system as a whole should be described as positive or negative.

Another important feature of the paper of Jacob *et al* was their suggestion that specific sites on the cytoplasmic membrane might be important in controlling plasmid replication and segregation. If membrane attachment sites were essential for plasmid replication, the number of sites of a particular type could determine the number of copies of the corresponding plasmid which the cell could maintain. If different plasmids attached to the same site they would be incompatible because they would compete for a limited number of sites. Membrane sites could also be important in plasmid segregation. Following replication, daughter plasmids could remain attached to adjacent sites on the membrane; segregation of the two plasmids could then occur through membrane growth which moved the sites, and their attached plasmids, into daughter cells.

Evidence bearing on the replicator model The model for replicon control proposed by Jacob *et al* is referred to here as the 'replicator model', rather than the 'replicon model'; the term 'replicon' has since acquired a use which is independent of the model proposed to explain its replication.

In relation to plasmids, the key features of the replicator model as proposed by Jacob *et al* (1963) are assumed to be the following:

1 A positively-acting plasmid-specified product (initiator) initiates replication by acting as the replicator site of a plasmid which is attached to a specific site on the cytoplasmic membrane.

2 Incompatibility arises because some plasmids require the same membrane attachment site for replication.

3 Plasmids-segregate at cell division because they are attached to sites on the cytoplasmic membrane which move into daughter cells.

The mechanism for timing replication in relation to the cell growth rate was not specified by Jacob *et al.* It was suggested that plasmid replication might be initiated by a signal from the membrane. It might also begin when a new attachment/replication site was formed on the membrane or when a burst of initiator was produced.

It is interesting to look at how many of the recent results in fact support the replicator model. Certain plasmids do produce a positively acting initiator, for example the *Inc*FII plasmids like R1 produce the *rep*A1 protein. However replication as a whole is controlled by negative feedback: RNA I, which is synthesized constitutively, inhibits the formation of *rep*A1. When plasmids replicate, the concentration of RNA I increases momentarily to inhibit further replication.

The role of the cytoplasmic membrane in plasmid replication is still unclear. Incompatibility of plasmids such as ColE1 result from the mechanism which controls plasmid replication and plasmid copy number. It does not appear to arise from competition for membrane attachment sites. However, the incompabibility properties of the F plasmid appear to result, at least in part, from the mechanism which ensures plasmid segregation; this mechanism might involve membrane attachment.

Summary

Structure and isolation Bacterial plasmids are circular molecules of double-stranded DNA which can replicate and be inherited without being linked to the bacterial chromosome. Their molecular weights range from 1×10^6 to about 200×10^6.

Most of the plasmid molecules in bacteria are covalently closed circular (CCC) molecules which have superhelical twists. An open circular molecule lacking superhelical twists is formed when one of the two strands of a CCC plasmid is broken. CCC plasmids can be isolated from cell extracts by buoyant density centrifugation in the presence of the dye ethidium bromide.

Replication The two major groups of plasmids found in enterobacteria are small non-conjugative plasmids (molecular weight less than 10×10^6) which are maintained at about 15 copies per bacterial chromosome (example ColE1), and large conjugative plasmids which are maintained at one or two copies per bacterial chromosome.

ColE1 replication depends entirely on host enzymes. Replication begins at a fixed point, the origin, and continues in one direction away from the origin, i.e. replication is unidirectional. The first event appears to be the transcription of a region of the plasmid near the origin to form an RNA molecule called RNA II. Transcription is catalysed by the DNA-dependent RNA polymerase of the host and provides an RNA primer for subsequent DNA synthesis. DNA synthesis, catalysed initially by DNA polymerase I, begins at the 3′ OH end of the primer and continues in a 5′→3′ direction for about 500 nucleotides. Discontinuous replication of the other strand then begins. Short RNA primers (made by the *dnaG* protein) are extended by DNA polymerase III holoenzyme (in combination with several other proteins) to form Okazaki fragments of about 1000 bases. The RNA primers are replaced by DNA (DNA polymerase I catalyses this reaction) and adjacent DNA fragments are then joined by DNA ligase. Other proteins required for replication include single-strand DNA binding protein, which unwinds the double helix, and DNA gyrase, which introduces negative superhelical twists.

Many large conjugative plasmids, such as F and R1, which are found in enterobacteria, differ from ColE1 in that they can replicate in *polA1* mutants which have low levels of DNA polymerase I activity. These conjugative plasmids code for at least one protein which is essential for their replication and which appears to be required to initiate a round of replication. Almost all the host proteins required for replication of the *E.coli* chromosome are also necessary for replication of conjugative plasmids in *E.coli.*

The rate of plasmid replication in relation to cell growth is believed to be coordinated by a mechanism which controls the rate of initiation of replication. ColE1 replication is controlled by a negative feedback mechanism. An RNA molecule (RNA I), together with the *rom* (or *rop*) protein, which is also encoded by the plasmid, inhibit the action of RNA II.

The initiation of replication of plasmids R1, which belongs to the IncF II group, is also controlled by an RNA molecule. This binds to another plasmid-specified RNA which encodes a protein (the *rep*A1 protein) which is needed for replication.

The mechanisms which ensure that each daughter cell receives a copy of a newly replicated plasmid at cell division are not understood, but it is thought that plasmids become attached to sites on the cytoplasmic membrane which segregate into daughter cells.

The mechanisms controlling plasmid replication and segregation appear to be responsible for plasmid incompatibility. Bacteria often contain two or more different (compatible) plasmids which are stably inherited. Some plasmids are incompatible because they produce similar inhibitors which control their copy numbers. In addition, plasmid incompatibility might be caused, at least in part, by competition for essential components, perhaps on the cytoplasmic membrane, which are required for segregation into daughter cells at cell division. Plasmids found in enterobacteria are classified into about 25 incompatibility groups.

References

FREIFELDER, D. (1976). *Physical Biochemistry*. Freeman, San Francisco.

JACOB, F., BRENNER, S. and CUZIN, F. (1963). On the regulation of DNA replication in bacteria. *Cold Spring Harbor Symposium on Quantitative Biology* 28: 329–48.

KORNBERG, A. (1983). Mechanisms of replication of the *Escherichia coli* chromosome. *European Journal of Biochemistry*. 137: 377–382.

NORDSTRÖM, K., MOLIN, S. and LIGHT, J. (1984). Control of replication of bacterial plasmids: genetics, molecular biology, and physiology of the plasmid R1 system. *Plasmid* 12: 71–90.

SCOTT, J. R. (1984). Regulation of plasmid replication. *Microbiological Reviews* 48: 1–23.

TOMIZAWA, J-I. and SO, M. T. (1984). Control of ColE1 plasmid replication: enhancement of binding of RNA I to the primer transcript by the Rom protein. *Cell* 38: 871–878.

3 Conjugation

Many plasmids are conjugative. They have a cluster of genes which enables them to transfer copies of themselves from one bacterium to another by conjugation. Plasmids can also be transferred in ways that do not depend on plasmid-encoded products, that is, by bacteriophage-mediated transduction or by transformation. During bacteriophage infections, a plasmid is sometimes enclosed in a phage coat to form a transducing particle which can inject the plasmid into a suitable recipient. Plasmids can also be transferred by transformation. Some bacterial species can simply take up DNA molecules from solution, but others require special conditions (for example, high concentrations of $CaCl_2$). The extent to which transformation and transduction occur in nature is unknown, but conjugation is probably the most important means of plasmid dissemination.

Conjugative plasmids have been found in strains belonging to many different groups of Gram-negative bacteria, but only in four Gram-positive genera, *Streptococcus, Streptomyces, Clostridium* and *Bacillus*.

Many conjugative plasmids can also transfer copies of chromosomal genes between bacteria, although this usually occurs at a much lower frequency than transfer of the plasmids themselves. Plasmids usually transfer only fragments of chromosomal DNA; transfer of an entire chromosome is very rare. The frequency of chromosome transfer in nature is unknown, but plasmids presumably contribute to the formation of new combinations of chromosomal genes and hence affect the rate of bacterial evolution. Some plasmids can transfer themselves, and fragments of chromosome, between most types of Gram-negative bacteria. In a sense, therefore, most Gram-negative species share a common gene-pool because of the activities of these wide host-range plasmids.

The most extensively studied conjugative plasmid—and the first to be discovered—is the F plasmid of *E.coli* K-12. The ability of this plasmid to transfer chromosomal genes between strains of *E. coli* has been invaluable in constructing the genetic map of this bacterium. In contrast to many other plasmids, F does not specify any other obvious traits, such as a colicin or antibiotic-resistance, in addition to conjugative ability. The *E.coli* K-12 strain containing F has been grown in laboratory culture media since it was first isolated in California in the early 1920s. It would not be surprising if plasmid variants have been selected which are better adapted to laboratory conditions. This might account for some of the atypical features of F.

F and F-like plasmids

F is a circular DNA molecule with a molecular weight of 63×10^6. About a quarter of the plasmid, the *tra* region comprising about 21 genes, codes for proteins which are essential for transferring copies of F by conjugation (Fig. 6). The nature and functions of the proteins encoded by the rest of the plasmid are

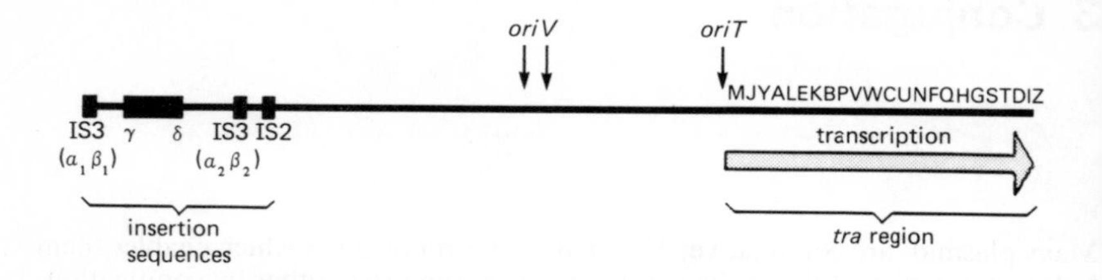

Fig. 6 The F plasmid. *oriT* is the origin of replication associated with conjugal transfer. Replication usually begins at the origin of vegetative replication (*oriV*) shown on the left as drawn, but it can begin at the other origin

largely unknown, but four *insertion sequences* are known to be present; these are involved in the conjugal transfer of genes from the bacterial chromosome.

F-like plasmids are plasmids whose *tra* regions are similar to that of the F-plasmid. Homology between many of the genes required for conjugation in F-like plasmids and the corresponding genes in F can be shown by genetic and physical techniques, including heteroduplex analysis. Many F-like plasmids specify colicins (Col plasmids) or antibiotic-resistance (R plasmids).

The most conspicuous features coded for by F and F-like plasmids are sex pili (Fig. 7). These are hollow tubes, composed largely of protein, which protrude from cells up to a length of 20μm (although most are about 2μm in length). They are essential for conjugation, although their role in this process is still not entirely clear.

The *tra* operon

Almost all the genes required for transfer of F plasmid DNA by conjugation are contained within a single operon, the *tra* operon. The structural genes within this operon are transcribed into a messenger RNA molecule comprising about 30 000 bases. The *tra* genes are all transcribed from the DNA strand which is transferred during conjugation.

Many of the *tra* genes are essential for the formation of sex pili, although the pili themselves comprise only one type of protein sub-unit. F plasmids with mutations in any of the following genes are unable to specify the synthesis of sex pili: *traJ, A, L, E, K, B, V, W, C, U, F, Q, H* and *G*. The protein in pili is specified by *traA*. The roles of the other gene-products in the formation of pili is unknown.

The products of *traY, traZ, traM, traI* and *traD* appear to be involved in the transfer of DNA; plasmids with mutations in any of these genes specify the synthesis of pili but are nevertheless unable to transfer copies of themselves by conjugation. The products of *traG* and *traN* appear to be involved in the formation of stable connexions between donors and recipients. F^+ bacteria normally adhere firmly to recipients in broth cultures. They may become attached to single bacteria or to groups, forming clumps or *mating aggregates*. However, if the donor harbours a *traG* or *traN* mutation the few clumps which form are more easily disrupted. The *traS* and *traT* gene-products are responsible for surface exclusion (p.34).

The first steps in producing the map of the *tra* operon were the isolation of *tra*

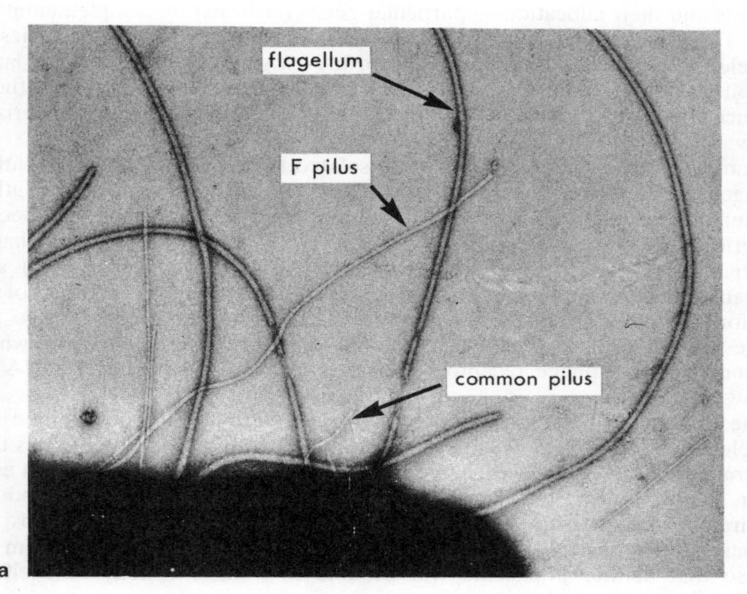

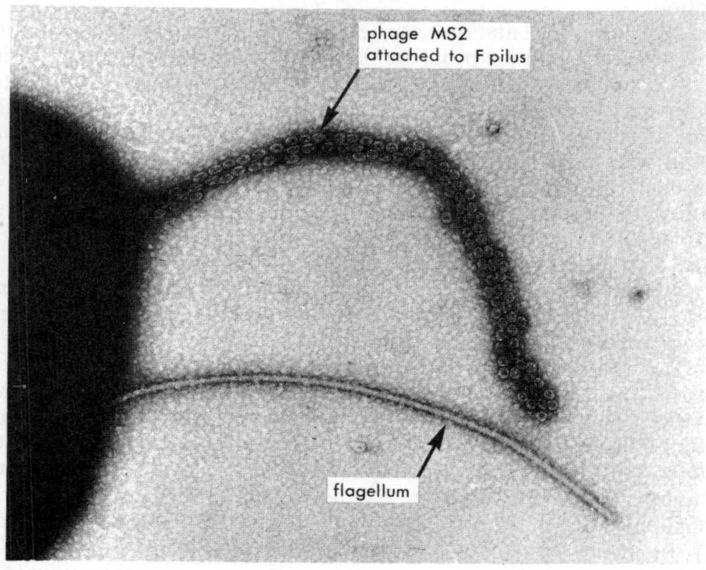

Fig. 7a *Escherichia coli* K-12 showing an F pilus (coded for by the F plasmid), flagella, and common pili; **b** Attachment of the F-specific bacteriophage MS2 to an F pilus. (Photographs kindly provided by A M Lawn)

mutants and their allocation to particular genes (cistrons) by complementation tests. The arrangement of the *tra* genes was determined from the properties of *tra* deletion mutants. The organization of the *tra* genes in the form of a large operon was determined from the effects of polar mutations, especially those produced by insertion of the bacteriophage Mu. The experiments are summarized below.

Transfer-deficient mutants of F were isolated by treating F^+ cultures with a mutagen and selecting those which had lost the ability to transfer F. Other mutants were isolated by selecting F^+ bacteria which were resistant to F-specific bacteriophages. F-specific phages are of two types: single-strand RNA phages which adsorb to the sides of sex pili (for example, MS2 and Qβ; see Fig. 7), and filamentous phages containing single-strand DNA which adsorb to the ends of sex pili (for example, f1). The F pilus is an essential receptor site for these phages, so phage-resistant mutants usually form no pili at all (or only defective pili) which are ineffective as phage receptors and are also unable to transfer DNA by conjugation.

The number of genes essential for conjugation was determined by complementation tests between *tra* mutants. The principle of these tests is that two *tra* mutants of F which are defective in *different* genes can complement each other. If both are present in the same cell, the complete set of gene-products required for conjugation will be formed. However, if the two plasmids are defective in the *same* gene, an essential *tra* gene-product will be missing from the cell so that neither F plasmid will be able to transfer a copy of itself by conjugation.

In practice, complementation tests between *tra* mutants of F must be carried out in ways which avoid the problem of incompatibility; two F plasmids cannot stably co-exist in the same cell (see p.15). In various experiments, four different procedures have been used to overcome the problem of incompatibility. The first complementation tests were made using R100.1, an F-like R plasmid which is compatible with F and has a *tra* operon very similar to that of F. Many of the *tra* gene-products specified by R100.1 are so similar to those of F that they can complement the defective products specified by F *tra* mutants. Ohtsubo defined six complementation groups from the results of tests between *tra* mutants of R100.1 and of F′*gal*$^+$ (for the structure of F′ plasmids see p. 42).

However, not all F *tra* genes can be complemented by the corresponding *tra* genes of R100.1; the products of some genes are highly specific and act only to promote transfer of the plasmid which codes for them (see, for example *traI*, below). The 'transient heterozygote' method was devised by Achtman to allow analyses of complementation between F *tra* mutants during a short period of co-existence before they segregate. Two further methods avoid the problem of incompatibility by joining fragments of the *tra* region to another replicon, either a plasmid or a phage, which is compatible with F. *In vitro* recombination procedures have been used to clone fragments of F using plasmid vectors which are compatible with F, and Willetts obtained λ transducing phages *in vivo* in which parts of the *tra* region were incorporated into the λ chromosome. It is sometimes necessary to transfer *tra* mutants of F between strains in order to carry out complementation tests. This can be done either by transduction of the plasmid by phage P1, by transferring amber *tra* mutants from a Su^+ (suppressor) strain to

a Su⁻ strain in which the amber mutation will not be supr
transformation.

The order of genes within the *tra* region was determine
complementation tests between deletion mutants and point mutants. D
mutants of autonomous F plasmids and of integrated F plasmids in Hfr stra
were used, as well as fragments of F which had been cloned into plasmid vectors or attached to phage chromosomes. The principle of these tests is that deletion mutants or fragments are tested for their ability to complement a variety of point mutants. Complementation indicates that a functional copy of the gene is present in the deletion mutant or cloned fragment. From the results obtained from a series of deletion mutants the order of genes can be determined.

From the analysis of polar mutations almost all the *tra* genes were found to lie within a single operon. Polar mutations not only affect the synthesis of the protein specified by the gene in which they occur, but also the production of all proteins specified by distal genes; that is, those *downstream* in the operon. Amber mutations have this affect, as does insertion of the bacteriophage Mu. The *traM* gene was found to lie outside the *tra* operon and insertion of Mu in *traD* did not completely prevent production of the *traI* gene-product; *traI* can be transcribed from a separate promoter although it also is part of the *tra* operon. The *traJ* gene is also excluded from the *tra* operon. This gene codes for a product which is required for synthesis of the proteins encoded by genes in the *tra* operon (see next section). The *traM* and *traJ* genes have their own promoters and are probably transcribed in the same direction as all the other *tra* genes.

Control of the *tra* operon

The genes required for the transfer of most conjugative plasmids are expressed in only a small proportion of the cells which harbor them. The conjugative ability of almost all F-like R plasmids isolated from enterobacteria is expressed in only about 0.1% of R^+ cells. In most R^+ cells, the genes comprising the *tra* operon are not transcribed into messenger RNA. The F plasmid and several F-like ColV plasmids are exceptional in that their *tra* operons are expressed constitutively; all F^+ cells produce a few sex pili (usually between 1 and 3) and can act as donors of plasmid DNA.

Regulation of the *tra* operon of F-like plasmids appears to be brought about by a repressor which acts at an operator to inhibit transcription of the operon. The *tra* operon is therefore regulated by a negative control system similar to that of the *lac* operon. Expression of the *lac* operon is controlled by a repressor, specified by *lacI,* which acts at an operator site on the DNA to inhibit transcription of the structural genes, *lacZ, lacY* and *lacA*. Control of the *tra* operon is not yet fully understood, and it may involve more elements than occur in the *lac* control system.

The essential features of the negative control system which represses the conjugative ability of most F-like plasmids were determined by analysing plasmid mutants which were de-repressed for conjugal transfer. De-repressed mutants express donor ability constitutively; all the cells in a populaton harbouring a de-repressed (*drd*) mutant of an F-like R plasmid are able to donate copies of the plasmid and are also sensitive to F-specific phages as they usually produce at least 10 sex pili per cell.

If conjugative ability in F-like plasmids is controlled negatively in a manner analogous to the *lac* operon, then *drd* mutants may be of two types: those which fail to produce an effective repressor and those which produce a repressor but are no longer sensitive to its effects because of a mutation in the operator. In the notation often used for the *lac* operon, the former type of *drd* mutant can be described as $i,^{-},o^{+}$ (i^{-} indicates that an effective repressor is not produced) and the latter as i^{-}, o^{c} (o^{c} means *operator-constitutive*). The two types can be distinguished by comparing their effects on the conjugative ability of other F-like plasmids. The predicted effects of de-repressed mutants on the conjugative ability of another plasmid present in the same cell are listed in Table 1. The important points are that i^{+},o^{c} plasmids are always *cis*-dominant, whereas i^{-},o^{+} plasmids may be either *trans*-dominant or recessive when in combination with a plasmid which is i^{+},o^{+}. (Plasmids of the dominant, i^{-d}, type would have mutations analogous to mutations in the *i* gene of the *lac* operon which are constitutive for expression of *lac* structural genes and are dominant in i^{+}/i^{-d} partial diploids. The i^{-d} mutations are believed to be dominant because the active form of the *lac* repressor is a tetramer; most of the tetramers present in partial diploids will be composed of two types of sub-unit, the presence of the altered polypeptide apparently preventing the hybrid tetramer from binding effectively to the operator.)

The difficulties encountered in performing the experiments to test these predictions are similar to those which arose in connexion with mapping *tra* genes, as described previously. The experiments cannot be carried out in the most straightforward way because two mutants derived from the same plasmid will be incompatible. The alternative approaches are therefore to analyse interactions between F-like plasmids belonging to different incompatibility groups, to analyse interactions between plasmids from the same incompatibility group during the short period of time before they segregate, or to clone the relevant fragments in plasmid vectors. Differences in the specificities of operators and repressors of different plasmids might affect the outcome of experiments if the first approach is used, while the results from the second approach might be affected by incompatibility reactions as well as by interactions between the *tra* control systems. The major findings from each of these three approaches are discussed below.

The effects of several F-like plasmids on the conjugative ability of F provided the first indication that the F-like group of plasmids had similar *tra* control systems. F belongs to incompatibility group FI and is therefore compatible with many F-like R plasmids and Col plasmids which belong to incompatibility groups FII, FIII, or FIV. Unlike F, almost all the Col plasmids and R plasmids belonging to these groups are naturally repressed, so that only about 1% to 0.1% of cells can act as donors. When these plasmids were transferred to an F^{+} strain they greatly reduced the conjugative ability of F and were said to cause *fertility inhibition*. They were described as being fi^{+} or fin^{+}. The extent of the fertility inhibition caused by different fi^{+} plasmids varied; some caused no more than a 20-fold decrease in the frequency of F transfer, whereas others reduced the frequency more than 10 000-fold. F^{+} cells which also contain an fi^{+} plasmid fail to produce F sex pili.

The ability of many fi^{+} F-like plasmids to repress the *tra* operon of F suggested that these plasmids specified repressors which not only repressed their own *tra* operons but also acted in *trans* to repress the F *tra* operon. F therefore appears to be a naturally-occurring plasmid of the i^-, o^+ type, i.e. it does not itself code for an active repressor although it has an operator which can be acted upon by repressors specified by other (fi^+) F-like plasmids. It is not known whether F specifies a repressor protein which is ineffective or no repressor protein at all.

The properties of several de-repressed R plasmids and Col plasmids (mainly members of incompatibility group FII) are consistent with the above interpretation. The *drd* mutants were classified into the groups i^+, o^c and i^-, o^+ from their interactions with F and with ColB-K98. The former type of mutant should be *cis*-dominant. It should therefore be unaffected by ColB-K98 (i^+, o^+; incompatibility group FIII) and should be able to repress F (i^-, o^+; incompatibility group FI). The latter type of mutant should be repressed by ColB-K98 (unless the i^- mutation was of the dominant type—see above, and Table 1) and should be unable to repress F. Numerous *drd* plasmid mutants fell unambiguously into one of the two groups when their interactions with ColB-K98 and with F were analysed. R plasmids caused various degrees of fertility inhibition, indicating that their repressors differed in their effectiveness for F.

These results are therefore consistent with control of the *tra* operon by a repressor acting as an operator. In the current nomeclature, the repressor is specified by gene *fin0* and the operator site is designated *tra0*.

Table 1 Possible interactions between repressed plasmids and their de-repressed mutants

Plasmid combination (genotypes)			Ability to transfer by conjugation
$i^+ o^c$	(*drd*)		de-repressed
$i^+ o^+$			repressed
$i^- o^+$	(*drd*)	*trans*-recessive	repressed
$i^+ o^+$			repressed
$i^{-d} o^+$	(*drd*)	*trans*-dominant†	de-repressed
$i^+ o^+$			de-repressed
$i^- o^+$	(*drd*)		de-repressed
$i^- o^+$	(*drd*)		de-repressed
$i^- o^+$	(*drd*)	intragenic complementation††	repressed
$i^- o^+$	(*drd*)		repressed

It is assumed that conjugal transfer is controlled by a repressor, specified by gene *i*, which acts at an operator, *o*, to inhibit transfer. Possible interactions between pairs of plasmids, either a *drd* mutant and a repressed (wild type) plasmid or two *drd* mutants, are listed.

† See text p. 28.

†† Intragenic complementation could occur if the active form of the *tra* repressor is composed of two or more sub-units of repressor protein. Combination of repressor proteins specified by different mutants which are inactive alone can be active.

An alternative approach to investigating the *tra* control system is to analyse interactions between incompatible plasmids during the short period of co-existence before they segregate. Cells containing mutations derived from the same plasmid are known as *transient heterozygotes*. Transient heterozygotes were used to analyse mutants of F$'$*lac*$^+$ which were able to transfer in the presence of the *fi*$^+$ R plasmid, R100 (incompatibility group, FII). On the simplest model, outlined above, all such mutants should either be of the o^c type or, possibly, mutants which were i^- (dominant) rather than i^- (recessive). When the mutants were examined in transient heterozygotes they did not behave as expected, as if they were o^c. Transient heterozygotes were made by transferring a mutant F$'$*lac*$^+$ (i.e. a mutant not subject to fertility inhibition by R100) into a strain already containing F$'$*his*$^+$ and R100. The donor of the mutant F$'$*lac*$^+$ was then killed with phage T6 and the transient heterozygote, which was T6-resistant, was mated with an F$^-$ recipient for 30 minutes. It was found that transfer of the F$'$*lac*$^+$ mutant plasmid from transient herterozygotes already harbouring F$'$*his* and R100 was inhibited. Possible explanations for this result are: 1 The mutant F$'$*lac*$^+$plasmids were o^c but were unable to transfer from the transient heterozygote because of an incompatibility reaction with the resident F$'$*his*$^+$.2 The F$'$*lac*$^+$ plasmids had mutated from i^- (recessive) to i^- (dominant). (The repressor specified by an incoming F$'$ plasmid does not become active within the 30 minutes mating period, as described below). 3 The resident F$'$*his*$^+$ determined a *trans*-acting product (P_F, determined by a *finP* gene) which was necessary for fertility inhibition *in addition* to the repressor specified by R100.If the F$'$*lac*$^+$ mutant did not specify the synthesis of an effective P_F, it would be able to transfer at high frequency from a cell harbouring R100 (if the R100 did not determine a product which had the same specificity as P_F), but not from a cell harbouring both R100 and an F$'$*his*$^+$ specifying an active P_F.

The model for *tra* operon control which is consistent with the third explanation is that F-like plasmids have two genes, *finO* and *finP*. These act together at an operator, *traO*, to inhibit transcription of the *tra* operon. It must also be proposed that the *finP* gene-product is more specific in its effects than the *finO* gene-product. Thus, the *finP* gene-product of R100, for example, does not act on F, nor *vice versa*. On this model, the F plasmid is de-repressed because it is *finO*$^-$,*finP*$^+$,*traO*$^+$.

Genetic evidence indicates that the effect of the repressors is to inhibit transcription of traJ. This effectively inhibits transcription of the entire *tra* operon because *traJ* appears to code for a product which is required to switch on transcription of this operon. The mechanism of action of the *traJ* gene product is not understood.

The *finP* gene, which enables F plasmids to transfer in the presence of R100 and another F plasmid, has been mapped between *traM* and *traJ*. A protein specified by *finP* has not yet been identified.

Lambda phages carrying the *finO* gene of the *fi*$^+$ plasmid R100 were analysed to determine the position of the *finO* gene. It was mapped to a position between the end of the *tra* operon and the origin of replication, i.e. it was at least 35,000 base-pairs away from the *traJ* gene.

If the *finO* gene controls the R100 *tra* operon it is perhaps surprising that it is so far away from the operator controlling *traJ*, it's presumed site of action according to the model described above. The products of the *traO* and *finP* genes have not been identified.

Some plasmids which are clearly unrelated to F also inhibit the ability of F to transmit copies of itself by conjugation. The mechanism of this effect is unknown, but it appears to be unrelated to the effects of *fi*$^+$ F-like plasmids because several of the plasmids inhibit transfer of o^c (*traO*$^-$) F plasmids which are not inhibited by *fi*$^+$ F-like plasmids.

Whatever the precise nature of the mechanism which prevents expression of the *tra* operon in most cells harbouring F-like plasmids, it does not become effective immediately upon entry of a plasmid into a recipient. Repression of conjugative ability is established slowly in cells which have just acquired a plasmid, and may require several generations to become fully effective. Recently-acquired plasmids can therefore re-transfer copies of themselves to other recipients as if they were *drd* mutants.

The slow rate at which repression of plasmid conjugative ability is established was discovered by Stocker, Smith and Ozeki (see Meynell, 1972). The plasmid used in their experiments, ColIb-P9, codes for I-pili as opposed to F-like pili, but repression of naturally-repressed F-like plasmids occurs at similar rates. De-repression of newly-acquired plasmids leads to their rapid spread in broth cultures composed of donors and recipients, even though, initially, only a small proportion of cells can act as donors. At an appropriate stage in such a culture, a high proportion of cells which have recently acquired a plasmid will be able to transfer a copy of it as though they harboured *drd* mutants. The cells in such a culture are said to be in the HFT (high frequency transfer) state until the LFT (low frequency transfer) state is resumed. In a typical experiment, Stocker *et al* found that 8 generations were required before the conjugative ability of cells in a ColI$^+$ HFT culture became fully repressed. Starting with an HFT culture in which 25% of cells had de-repressed plasmids, after three generations in broth (diluted at intervals to keep the cell density at between 10^4 and 10^5 cells ml^{-1}) 11% of cells could act as donors, and after 8 generations only 0.01% of cells could donate plasmid copies.

Many operons are induced under appropriate conditions so that the bacterium can adapt to changes in its environment. Specific inducers of the *tra* operon which are comparable to lactose as an inducer of the *lac* operon have not been found. But two treatments which greatly increase the number of F-like pili produced by cells are rapid washing and the addition of anti-pilus antibody. The mechanisms of these effects, however, are unknown. The intracellular concentration of 3′ 5′ cyclic AMP also greatly affects the numbers of sex pili produced. Mutants which are either *cya*$^-$ (lacking adenylate cyclase activity and unable to synthesize 3′ 5′ cyclic AMP) or *crp*$^-$ (lacking cyclic AMP receptor protein) produce between 50 and 100 times the usual number of F pili per cell. The addition of 3′ 5′ cyclic AMP to cultures of *cya*$^-$ bacteria reduces the numbers of sex pili per cell to those produced by *cya*$^+$ bacteria under similar conditions. Perhaps the increased pilus production which occurs after cells have been washed is caused by a depletion of intracellular 3′ 5′ cyclic AMP.

Possibly, the de-repression of the *tra* operon which occurs for several generations after a plasmid has been transferred to a recipient (giving rise to the HFT state) is an adaptive response which has evolved because it enables plasmids to spread more efficiently. It could be argued that if a plasmid has found one suitable recipient, it is more likely to find others nearby (perhaps members of the same clone of cells) so that it is worthwhile retaining the de-repressed state for a few generations. But it is not known whether the kinetics of de-repression is really an evolved adaptation, or simply a consequence of the nature of the *tra* control mechanism. Bacteria carrying certain streptococcal plasmids are induced to mate by small peptides released from potential recipient cells, as described below.

De-repressed plasmid mutants can readily be isolated in the laboratory, but it appears that such plasmids are at a selective disadvantage in natural environments. Probably the most important factor which prevents the spread of de-repressed plasmids in nature is that they greatly decrease the growth rate of their bacterial hosts. This is very evident when the colonies of cells containing de-repressed plasmids are compared with those containing the naturally-repressed form of the plasmids. De-repressed mutants invariably reduce colony size on solid media and also reduce the growth rate of bacteria in liquid cultures. The diversion of cell resources to the production of *tra*-specified proteins presumably accounts for much of the decreased growth rate, but de-repressed plasmids also weaken the cell envelope. Cells containing them are more easily lysed by detergents than are cells containing repressed plasmids. The reason for this is unclear. Cells containing de-repressed plasmids are also at a disadvantage because they are sensitive to bacteriophages which depend on plasmid-specified sex pili for infection.

Sex pili

The most extensively studied types of sex pili are the F-like and I-like types, pili closely resembling those specified by F and ColI respectively.

Several lines of evidence indicate that sex pili have an essential role in conjugation. Most of these experiments have been carried out with cells harbouring F or F-like plasmids. Cells containing conjugative plasmids are unable to transfer DNA if they fail to produce sex pili, either because of a mutation in a *tra* gene or because expression of the *tra* operon is repressed. When sex pili are removed by shearing, the cells are unable to act as donors until new pili are formed on the surface. Attachment of antibody molecules or of bacteriophages to sex pili also inhibitis conjugation. F pili are composed of a single type of sub-unit which comprises a polypeptide, one D-glucose residue and two phosphate groups. The *traA* gene of the F plasmid encodes a protein (propilin) of 121 amino acids. This is cleaved to form the protein of 70 amino acids which is found in pili. The *traQ* gene product is involved in the cleavage of the propilin molecule.

Purified preparations of F pili retain many of the properties of pili still attached to bacterial cells. These preparations adsorb to F^- bacteria, and pilus-specific bacteriophages bind to them. Optical and X-ray diffraction techniques to determine the arrangement of sub-units have established that F pili are hollow tubes of about 8 nm diameter, the central hole having a diameter of about 2nm. The sub-units are arranged in the form of helices.

F-like pili specified by Col plasmids and R plasmids are all very similar, but they can be divided into four groups according to their immunological cross-reactions using sera raised against piliated bacteria. The pili specified by ColV, I-K94 are indistinguishable from those of F. Bacteria containing two different F-like plasmids form *mixed pili,* composed of two types of sub-unit. Such pili are perfectly functional and both types of plasmid can be transferred from cells which produce them.

I-like conjugative plasmids, which are frequently found in strains of *Escherichia coli, Shigella* and *Salmonella,* code for pili which rarely exceed 2μm in length and are a little thinner than F-like pili. ColIb-P9 codes for the prototype I pilus. Electron microscopy of I-pili indicates that they also have a hole through the central axis. There are no antigenic cross reactions between F-pili and I-pili; and I-pili adsorb bacteriophages which do not adsorb to F-pili. Only filamentous bacteriphages (If1 and If2) with single-strand DNA chromosomes have been found to adsorb specifically to I-pili; no I-specific isometric phages similar to the F-specific bacteriophages MS2 or Qβ have been found. If1 and If2 are serologically related to the F-specific filamentous phage M13.

Pili are also specified by R plasmids belonging to the incompatibility groups N, P, T, W and X. Evidence which indicates that the pili have an essential role in conjugation has been obtained for plasmids in groups I, N, P, W and X. Bacteriophages have been discovered which adsorb specifically to pili coded for by plasmids of groups N, P, T and W. Pili have not been seen on cells of Gram-positive bacteria carrying conjugative plasmids, suggesting that their mechanism of conjugation is fundamentally different to that of Gram-negatives.

Role of sex pili in conjugation Although several types of sex pili are essential for conjugation, their role in the process is not yet understood. The two most likely functions are: 1 to provide a hollow tube through which DNA passes from donor to recipient or, 2 to bring donor and recipient together by retracting once the tip of the pilus has become attached to the recipient. These two roles need not be mutually exclusive; the pilus might retract and then form a short channel where the two cell envelopes are close together.

Microscopic examination of pairs of mating bacteria provide some support for the hypothesis that pili act as conjugation channels through which plasmids can pass in the absence of extensive pilus contraction or prolonged contact of cell walls. Recipients which appeared to be only loosely connected to donors (presumably connected by pili) were found to have received DNA from the donors when they were separated by micromanipulation and allowed to form colonies of cells. The central canal of the F pilus would be wide enough to allow passage of a single-stranded molecule of DNA (the form in which DNA is transferred by conjugation—see below). F pili seem to disappear under certain conditions (for example, after adsorption of filamentous bacteriophages), suggesting that they can depolymerize at their bases and retract. These observations suggested that pilus contraction might be involved in conjugation, but other explanations have been proposed to account for these results.

Whatever the precise role of the sex pilus in transferring plasmid DNA, it must be able to recognize suitable recipient cells and form stable connexions with them. Presumably, it is the tip of the pilus which is important in this respect. The F pilus forms stable connexions with *E.coli* but not with *Pseudomonas* or *Bacillus.* This suggests that there are receptors in the *E.coli* cell wall to which the pilus binds. However, the site to which the pilus binds appears to be far less specific than the receptor sites for bacteriophages or for colicins (see p.89), which can usually be identified as particular proteins or lipopolysaccharide structures in the envelope.

Mutations in the *ompA* gene, which codes for an outer membrane protein, or mutations which affect the lipopolysaccharide composition of the cell wall can greatly reduce the ability of cells to act as recipients in matings with F^+ donors.

F^+ donors are unable to form stable contacts with these mutants in broth; the relatively stable clumps of cells—*mating aggregates*—are absent from the cultures. However, although very little plasmid transfer occurs in broth, mating on memrane surfaces is unimpaired. So it appears that the altered cell surface in these mutants prevents the formation of stable connexions (and hence plasmid transfer) in broth, but if it is in fact a pilus receptor which is altered then it is not essential for conjugation in all conditions.

Surface exclusion Almost all conjugative plasmids specifying F-like or I-like pili are expressed in only a small proportion (about one cell in 10 000) of their host cells. The few cells which contain de-repressed plasmids synthesize pili and other proteins coded for by the *tra* operon which are essential for plasmid transfer. In addition, de-repression of the *tra* operon allows expression of two genes, *traS* and *traT*, which are not essential for plasmid transfer, but are apparently necessary to ensure that a cell does not attempt to mate with itself. This preventive mechanism is known as *surface exclusion.* Sex pili (especially F pili) are long, flexible appendages; the tip of a pilus might therefore become attached to a cell from which it was extruded and initiate plasmid transfer back into the same cell.

The proteins specified by *traS* and by *traT* appear to have evolved to prevent this happening. Cultures of bacteria which have constitutive expression of the *tra* operon do not continually transfer plasmids to other members of the same clone because the *traS* and *traT* genes are also de-repressed as part of the *tra* operon. However, it must be emphasized that constitutive expression of the *tra* operon is exceptional amongst F-like and I-like plasmids (and probably in most plasmids). Plasmids are expressed in only a small proportion of host cells, but these few cells will transfer plasmid copies into cells which already have (repressed) copies of the same plasmid, since *traS* and *traT* will be repressed in these recipients. Molecules of *tra* repressor protein in the recipients will prevent further expression of the incoming plasmid's *tra* operon, and incompatibility reactions will prevent its stable maintenance.

The mechanism of surface exclusion is not understood, but either the *traS* gene-product or the *traT* gene-product can act alone to bring about partial surface exclusion. The protein specified by *traT* is found in the outer membrane of F^+ bacteria and it prevents stable connexions forming between donors and *recipients. There are about 20,000 copies of the traT* protein in the membrane of an F^+ cell. It has been suggested that the *traT* gene-product might bind to the tip of the pilus so that it cannot bind to the recipient. The protein specified by *traS* occurs in the inner membrane; it prevents DNA transfer even though stable connexions between donors and recipients are formed.

Different F-like plasmids sometimes exclude each other, and presumably specify similar *traS* or *traT* gene-products. F-like plasmids can be classified into four surface exclusion groups on this basis.

Although F^+ cultures are poor recipients when mated with F^+ (or Hfr) donors, the surface exclusion barrier can be overcome by leaving them in a medium lacking nutrients. These starved cells can act as recipients when mated with F^+ donors and are referred to as *F^+ phenocopies.*

DNA transfer This topic has recently been reviewed by Willetts and Wilkins

(1984). Once effective contact has been made between donor and recipient, a signal is presumably transmitted to the recipient to initiate plasmid replication and transfer. The nature of the signal is unknown. Jacob *et al* (1963) proposed a key role for the membrane in plasmid replication. The high frequency with which the F plasmid was transferred once contacts were made between donors and recipients implied that the F plasmid was already attached to the cytoplasmic membrane at the base of the pilus. They proposed that a signal was in some way transmitted to the membrane site from a pilus which had made contact with a recipient. (Signals at these sites might also initiate replication not concerned with plasmid transfer by conjugation, but simply with the maintenance of plasmids in cultures of dividing cells, see p. 14) Electron microscopy indicates that F pili are formed at sites where inner and outer membranes of the cell envelope are in contact, but it is not known whether F plasmids are also attached to these sites.

The DNA which is transferred by conjugation (when conjugation is promoted by the F plasmid) is not a double helix, but a single-stranded molecule. DNA replication in both the donor and recipient accompanies the process of DNA transfer. However, replication is not essential for transfer; it can be inhibited without preventing plasmid transfer. The single-stranded F DNA is always transferred 5′-end first and it is always the *heavy* stand (the denser strand in poly (U,G)-caesium chloride gradients) which is transferred. The lighter strand stays in the donor cell. A 5′ end is peeled off the plasmid double helix and is transferred into the recipient. The single-stranded DNA remaining in the donor is the template for the synthesis of a complementary strand so that a plasmid double-helix is reformed. A plasmid double helix is also formed in the recipient by the synthesis of a polynucleotide complementary to the transferred strand.

The origin of DNA transfer *oriT* (which is, presumably, also the origin of replication associated with DNA transfer) has been mapped on the F plasmid. It is close to the origin of transcription of the *tra* operon and far removed from the origin of vegetative replication (Fig. 6). The nucleotide sequences of the *oriT* regions of F and of several other plasmids have been determined.

The properties of *traI, traM, traY* and *traZ* mutants suggest that these genes are involved in the transfer of plasmid DNA rather than, for example, in the formation of F pili. F pili are formed by cells containing these *tra* mutants, but plasmid transfer does not occur. Furthermore, cells containing the mutants can be infected by F-specific phages and can transfer copies of other plasmids (such as the non-conjugative plasmid ColEl) to recipients by F-mediated mobilization (p.50). Initiation of plasmid DNA synthesis, which normally takes place shortly after donors form stable connexions with recipients, does not occur in donors harbouring these *tra* mutants. The plasmid DNA remains closed-circular (normally, one strand is nicked when donors are mated with recipients). The products of *traY* and *traZ,* are required to nick F DNA at the origin of transfer, *oriT*. They might be endonucleases. The *traY* gene-product is found in the cell envelope and the *traZ* product occurs in the cytoplasm. The *traM* gene product is also required for plasmid transfer and might have a regulatory role in the process. The *traM* gene-product is found in the cytoplasmic membrane. The *traY* and *traZ* genes are at either end of the *tra* operon (Fig. 6); *traY* is the first structural gene to be transcribed and *traZ* is the last. The *traM* gene, which can be transcribed independently of the *tra* operon, maps close to *traY*.

Once the F plasmid has been nicked at oriT, the two DNA strands are unwound by the *traI* gene product, a DNA helicase, using energy from ATP. The *traI* gene product is found in the cytoplasm.

One strand is then transferred to the recipient cell while the other remains in the donor (Fig. 8). The single-stranded DNA in both the donor and recipient cells serve as a template for DNA synthesis so that double-stranded F plasmid DNA is produced in both cells.

DNA synthesis in the donor cell involves the host's DNA polymerase III holoenzyme (see p. 9), but details of the reaction which provides an RNA primer for initiation of DNA synthesis are unclear. DNA polymerase III holoenzyme is presumably also responsible for synthesis of the complementary strand in the recipient cell.

Some plasmids such as F, depend on host enzymes for the synthesis of the RNA primer necessary to initiate DNA synthesis, but others, such as ColI (see p. 31) encode a primase. It appears that the enzyme itself is transferred along with the DNA. The RNA primers might be synthesized in the recipient cell by the transferred primase or they might already be synthesized in the donor for transfer along with the donated DNA. Once a complete copy of the plasmid has been transferred the two ends must be ligated at *oriT*. It has been suggested that the F-plasmid encoded endonuclease, the *traY* and *traZ* gene products, which nick at *oriT*, might be transferred along with the DNA and bring about a religation of 5′ and 3′ ends in the recipient.

Perhaps one of the most remarkable features of the conjugation process is the ability of the donor to transfer such a large macromolecule through the cytoplasmic membrane of another cell; the hole in the membrane must be

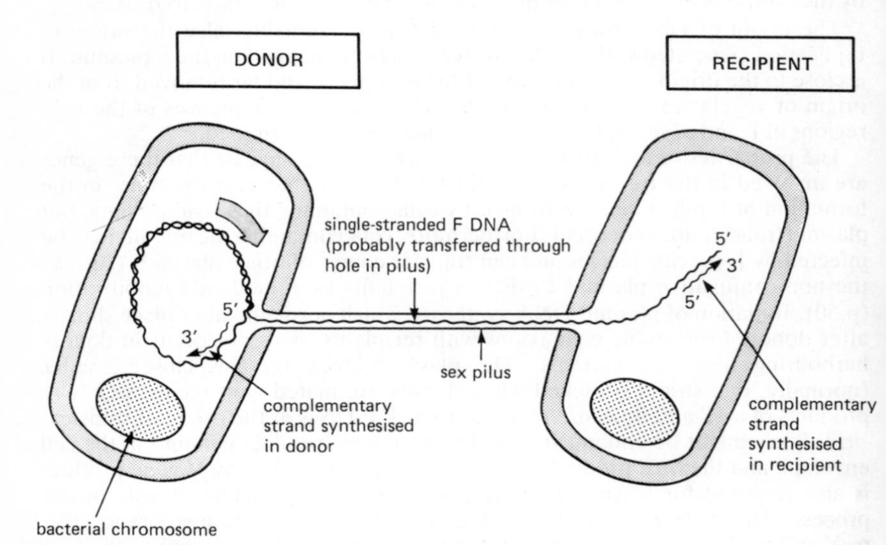

Fig. 8 Transfer of plasmid DNA during conjugation

exceptionally well sealed to prevent the entry and exit of molecules and ions which would otherwise lead to the death of the recipient. Bacteriophages have evolved similar abilities to transfer DNA through the cell envelope of bacteria.

Transfer of DNA from Hfr donors can continue for up to 100 min without killing the recipient. However, at high donor: recipient ratios, some recipients are killed (*lethal zygosis*) and macromolecules are released from them, so it appears that the recipient cannot maintain the integrity of its envelope if too many pili become attached. The nature of the channel in the recipient envelope through which DNA is passed without allowing intracellular components to escape is unknown. Perhaps the tip of the pilus fuses with the recipient cytoplasmic membrane to create a suitable channel.

Formation of Hfr strains

The latest genetic map of the circular chromosome of *E.coli* K-12 lists the positions of more than 1000 genes and genetic loci. Most of these were mapped by making use of the ability of F to transfer chromosomal genes between strains.

Two features of the F plasmid which make it particularly useful for analysing the genetic structure of *E.coli* K-12 are its ability to form Hfr strains and its ability to form F′ (F-prime) plasmids. In addition, because the *tra* operon of F is de-repressed, it transfers both itself and chromosomal genes at a higher frequency than most other plasmids. In Hfr strains, the F plasmid has recombined with the chromosome of *E. coli* so that the bacterial chromosome has a copy of the F plasmid included within it; the F plasmid is no longer an autonomous genetic element. (Genetic elements which can exist in either autonomous or integrated forms are sometimes call *episomes.*) The *tra* genes are still expressed in Hfr strains. As the plasmid is physically attached to the bacterial chromsome, it can not only transfer a copy of itself but can also transfer copies of chromosomal genes by conjugation.

The second useful feature of the F plasmid is that pieces of the *E.coli* chromosome can be added to it to form F′ plasmids. These are particularly useful for complementation tests and for analysing dominance/recessive relationships between alleles since *E.coli* can be made diploid for regions of its chromosome by transferring an appropriate F′ into the strain.

The low frequency transfer of chromosomal genes by an autonomous F plasmid was discovered before the much higher frequency of transfer by Hfr strains. However, in describing the transfer of chromosomal genes by conjugation, we will begin with a description of transfer by Hfr strains since this has been the most useful mode of transfer for studies on genetic mapping. Also, more is known about the mechanism of chromosome transfer by Hfr strains than about the transfer of chromosomal genes promoted by autonomous F plasmids.

Insertion sequences

Integration of F to form an Hfr can be represented as a recombination event involving a breakage and rejoining of plasmid and chromosomal DNA. Analysis of the plasmid and chromosomal sites involved in forming ten different Hfr strains indicates that most are formed by recombination between a homologous *insertion sequence* present both plasmid and chromosome (see Davidson *et al.*, 1975). Recombination at insertion sequences (or IS-elements) to form Hfrs can occur in *recA* strains. These sequences may not therefore simply provide sites of homology at which recombination can occur; they are special sequences at which recombination and other phenomena occur in the absence of host recombination mechanisms. Indeed, many important features of plasmids can be understood only with reference to the properties of the insertion sequences which they contain.

Insertion sequences are genetic elements of between 768 and about 5700 base pairs. Some of the important characteristics of the seven kinds of IS-element found in the *E.coli* K-12 chromosome are listed in Table 2. Several of these elements have also be found in the chromosomes of many bacterial species in addition to *E.coli,* as well as in a variety of bacteriophage and plasmid replicons. A search for insertion sequences in species other than *E.coli* has also revealed other types; more than 15 different kinds of IS-elements have been found.

Insertion sequences are closely related to *transposons*. The essential difference between transposons and insertion sequences is that the former specifies a readily observable phenotype, such as antibiotic-resistance. Transposons seem to be of two main types. *Composite transposons* (see Grindley and Reed, 1985) comprise two identical IS-elements which flank a gene or genes specifying, for example, resistance to a particular antibiotic. Thus, two copies of IS*1,* IS*10,* IS*50* and IS*903* form the flanking regions of composite transposons Tn*9,* Tn*10,* Tn*5* and Tn*903* respectively, which encode resistance to various antibiotics (Table 4). In Tn*9,* the two copies of IS*1* are in the same orientation: in the other transposons, the flanking IS-element form inverted repeats. In composite transposons, the flanking IS-elements are responsible for transposition of the element as a whole and the central part for specifying a trait such as antibiotic resistance. The second type of transposon is represented by Tn*3* and is not comprised of discrete IS-elements. Tn*3* is described in more detail in Chapter 4. The $\gamma\ \delta$ insertion sequence has a similar structure to Tn*3*.

Insertion sequences were first recognized from their ability to cause mutations by becoming inserted into genes of *E.coli* or of bacteriophage λ. They are elements capable of *transposition,* that is, they can become inserted into DNA molecules with which they have little or no base sequence homology. Most IS-elements can become inserted into any *E.coli* gene or into any replicon which may be present in *E.coli.* Insertion sequences IS*1* or IS*2* can insert at many different sites, but they become preferentially transposed to certain regions of a DNA molecule. IS*1* preferentially transposes to sites which are rich in adenine and thymidine residues and within such regions it tends to insert at sequences which are partially homologous to the ends of IS*1*. Other insertion sequences, such as IS*4,* appear to be highly specific, and are repeatedly found to insert at exactly the same point in a gene.

Replication associated with transposition. For some time after they were discovered, it was unclear whether insertion sequences simply moved from one site to another, or whether a *copy* of the insertion sequence was transposed to another site, the parent IS-element remaining at the original site. The evidence (summarised by Grindley and Reed, 1985), suggests that transposition of

IS-elements IS*1*, IS*10*, IS*50* and IS*903*(those which have been studied in most detail) does not usually involve the transfer of a plasmid copy. That is, it appears that an IS-element simply excises from the donor replicon, thus destroying it, and then becomes inserted at the new site. Only very occasionally (in less than 5% of transpositions) is a copy of the insertion sequence transposed. In contrast, the transposition Tn*3* and of related transposons is a replicative process; a copy of the transposon remains on the donor replicon molecule, which survives, and a copy is transferred to the recipient site. However, the transposition of both IS-elements and of Tn*3* is independent of the bacterial *recA* $^+$ gene product. (Recombination between homologous sequences almost never occurs in *recA* mutants). The transposable elements themselves encode proteins (transposases) which bring about their transposition.

IS-elements IS*10*, IS*50* and IS*903* apparently encode a single transposase. IS*1* seems to have two genes coding for proteins needed for transposition and IS*50* apparently encodes an inhibitor of transposition as well as a transposase. The transposases encoded by IS-elements are much less active in *trans* than when acting in *cis*. That is, a transposase is active on the IS-element which produces it but does not act effectively on a mutant IS-element in the same cell which cannot synthesize its owns transposase. This might be because the transposases, which are all highly basic proteins, rapidly bind to the DNA which encode them and then diffuse along the molecule until they reach the inverted repeats at the ends. The transpose specified by Tn*3* (see Chapter 4, p. 58) specifically recognizes the terminal inverted repeats of Tn*3;* the specific binding is dependent of ATP. Presumably, this is also the case for other transposases.

Terminal inverted repeat. A feature common to all insertion sequences and transposons (with only one exception, Tn*554*) is the presence of terminal inverted repeat sequences. That is, the same sequence (or very nearly the same sequence) is present at either end of the transposable element but in opposite orientations. For example, 18 out of 23 bases at one end of IS*1* are the same as those at the other

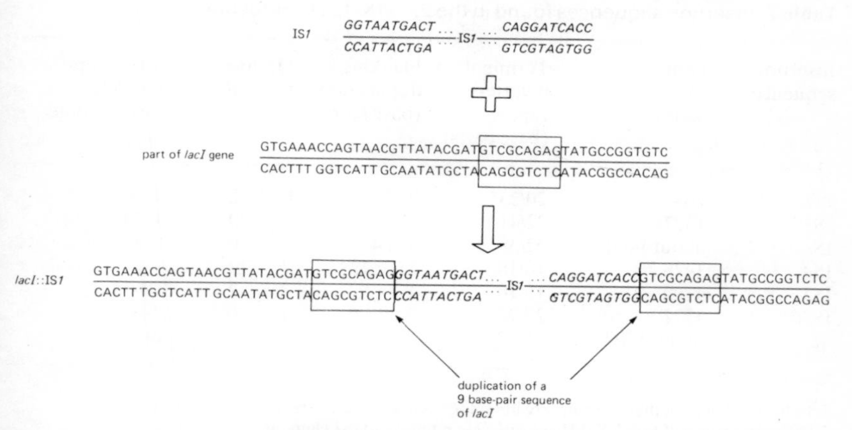

Fig. 9 Insertion of IS1 causes a 9 base-pair duplication. (From Calos, M. P., Johnsrud, L. and Miller, J. H. (1978). *Cell* 13: 411–18)

end, except that they are in an inverted order. Part of the terminal inverted repeat of IS*1* is shown at the top of Fig. 9. The inverted repeat sequences of six IS*1* elements from various sources are identical.

Duplication generated at insertion site. When a transposable element becomes inserted into a gene, the nucleotide sequence at the site of insertion (or 'target site') is duplicated (Fig. 9). As a result, the IS-element becomes flanked by a 2 to 13 base pair duplication. Tn*554* is exceptional and does not generate a duplication at the site of insertion.

The exact length of the duplication depends on the IS-element (Table 2). Duplication of a sequence at a target site suggests that during transposition a nuclease (presumably, the plasmid-encoded transposase) cuts the target DNA to form staggered cuts which are 3 to 12 base pairs apart on opposite strands of the double helix. (The type of cut envisaged would be similar to that produced by the restriction endonuclease *Eco*RI (see Table 8). Thus the simplest model for the insertion of a transposon is that the transposase recognises the ends of the IS-element, cuts it out and then joins it to the single-stranded ends at the target site. The single stranded DNA forms the template for DNA synthesis thus forming the flanking duplication characteristic of IS-element insertions. Another possibility is that simple insertion occurs by a modified form of a model proposed for replicative transposition, as described in Chapter 4 (Fig. 17).

DNA rearrangements. Insertion sequences and transposons can cause deletions and inversions of adjacent DNA sequences. The deletions caused by insertion sequences can be extensive—several genes adjacent to an IS-element can be deleted. Deletions of adjacent sequences may be accompanied by loss of the IS-element itself, but sometimes the sequence remains and can cause further deletions. Whether or not an insertion sequence is deleted together with adjacent sequences presumably depends on which of the two ends of insertion sequence generated the deletion.

Table 2 Insertion sequences found in the *E.coli* K-12 chromosone

Insertion sequence	Length (base pairs)	Terminal inverted repeat (base pairs)	Flanking duplication (base pairs)	Occurence F	Occurence RI	No. of copies *E.coli* K–12 chromosome
IS*1*	768	20/23⁺	9	0	2	4–10*
IS2	1327	32/41	5	1	0	4–13
IS*3*	about 1400	32/38	3 or 4	2	0	5–6
IS*4*	1426	16/18	11 or 12	?	?	1–2
IS*5*	1195	15/16	4	?	?	10–11
IS*30*	1221	23/26	2	?	0	2–8
$\gamma\delta$	about 5700	36/37	5	1	?	0–3

⁺ Indicates that 20 of the base pairs are homologous in an inverted repeat of 23 base pairs.
* Different strains of *E.coli* K-12 have different numbers of IS*1* elements.

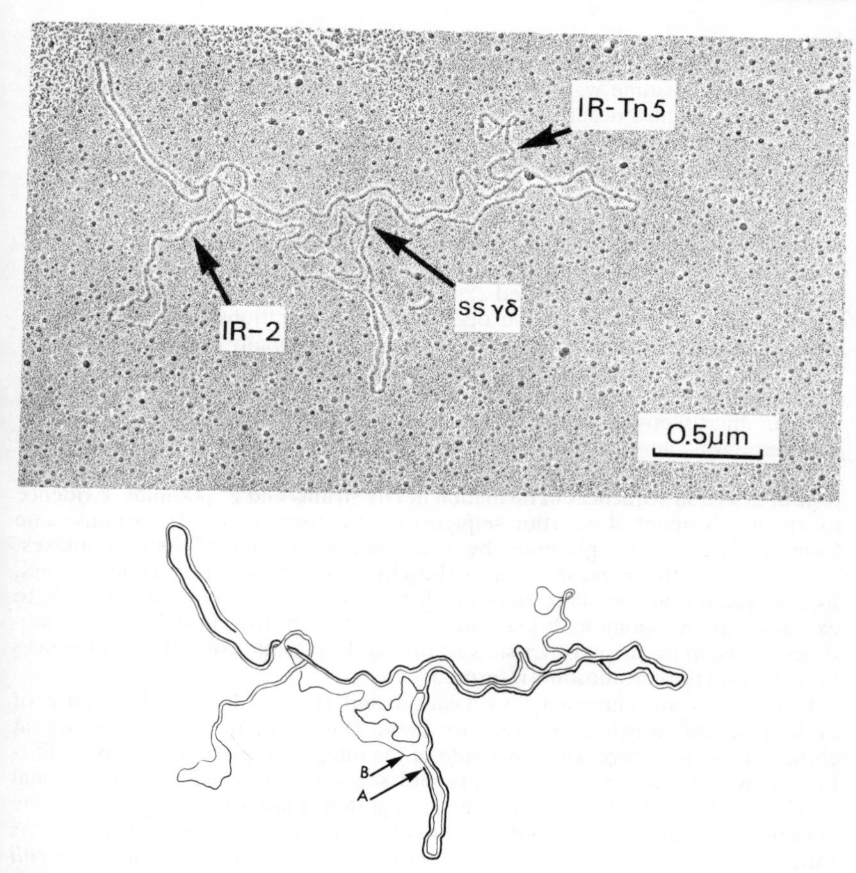

Fig. 10 Electron micrograph illustrating an insertion sequence (γδ), a transposon (Tn5), and inverted repeat sequences. The Tn*5* transposon and the γδ insertion sequence were transposed onto plasmid A to form plasmid B. A single-stranded molecule from B was then annealed with a fragment of A, forming a heteroduplex. Tn*5* has a long inverted repeat sequence at its ends; γδ has only a short terminal inverted repeat. Molecule B has another inverted repeat sequence, IR-2. Inverted repeats form a double-stranded stalk because of the complementarity of the bases forming the inverted repeat. From Davies, D. L., Binns, M. M., and Hardy, K. G. (1982) *Plasmid* 8:55–72.

An insertion sequence can become deleted from a DNA molecule so that the normal sequence of the gene, which was interrupted by the IS-element, is restored. Deletions of IS-elements which restore the original sequence occur infrequently, at a rate of 10^{-7} to 5×10^{-9}/cell/generation.

In addition to inactivating genes by insertion, causing deletions and other types of rearrangement, insertion sequences can also bring about more subtle changes in phenotype by modifying gene expression. Insertion of an IS-element sometimes enchances the expression of neighbouring genes; it may itself provide a promoter or in some way activate an existing promoter. Rearrangements of DNA sequences also occur within IS-elements. For example, a rearrangement within IS*2* can form new promoter sites; in one derivative of IS*2*, a 17 pair duplication of part of the IS-element created a new RNA polymerase binding site.

An IS-element may also provide a terminator of transcription. In fact, insertion sequences were first recognised from their ability to cause highly polar mutations in the *gal* operon; they not only inactivated the gene into which they were inserted, but also prevented expression of genes *downstream* in the same operon (distal in relation to the operator). The mutations were analysed by forming λ*gal* transducing phages (the *gal* genes occasionally become attached to phage lambda DNA as they are close to the lambda attachment site on the chromosome). Analysis of the transducing phage, using such techniques as electron microscopy of heteroduplexes, showed that the novel mutation in the *gal* operon was caused by an extra piece of DNA, an insertion sequence.

Role of insertion sequences in formation of Hfr strains and F′ plasmids Evidence for the involvement of insertion sequences in the formation of Hfr strains came from analyses of F′ plasmids by electron microscopy of heteroduplexes. Integration of the F plasmids into the chromosome is a reversible process. Recombination at the same sites which were involved in integration leads to excision—an autonomous F plasmid is released from the chromosome. Sometimes, excision occurs at other sites so that an F′ plasmid comprising sequences derived from the chromosome is formed.

F′ plasmids are classified into Types I and II according to the nature of excision event which gives rise to them (Fig 11). If excision occurs at chromosomal sequences on either side of the integrated plasmid, a Type II F′ is formed which has a complete F plasmid sequence attached to chromosomal sequences from both sides of the integrated plasmid. If a site on the chromosome and the F plasmid are involved in excision, then a Type I F′ is formed. In this case, part of the F plasmid sequence remains in the *E.coli* chromosome (to form a *sex factor affinity* locus, as described below) and the F′ has a corresponding deletion of part of the F plasmid sequence. Type I F′ plasmids are further classified into Types IA and IB (see Low, 1972). Type IA F′ plasmids have proximal genes of the parental Hfr (that is, the first chromosomal genes to be transferred by the Hfr), whereas Type IB F′ plasmids have distal genes (the last to be transferred).

F′ plasmids of Types I and II can be distinguished physically by the nature of the heteroduplex they form with F. Some can also be distinguished genetically according to whether they transfer both proximal and distal genes of the parental Hfr. The principle of the heteroduplex technique is that two DNA molecules are compared by denaturing each molecule to single strands of DNA and then allowing duplexes to form which are composed of polynucleotide strands from different molecules of DNA. To form plasmid heteroduplexes, open circular molecules of the two types of plasmid to be compared, say F and an F′, and first denatured by alkali or by heating so that each is separated into its two component strands. (Open circular molecules which have only one nick are

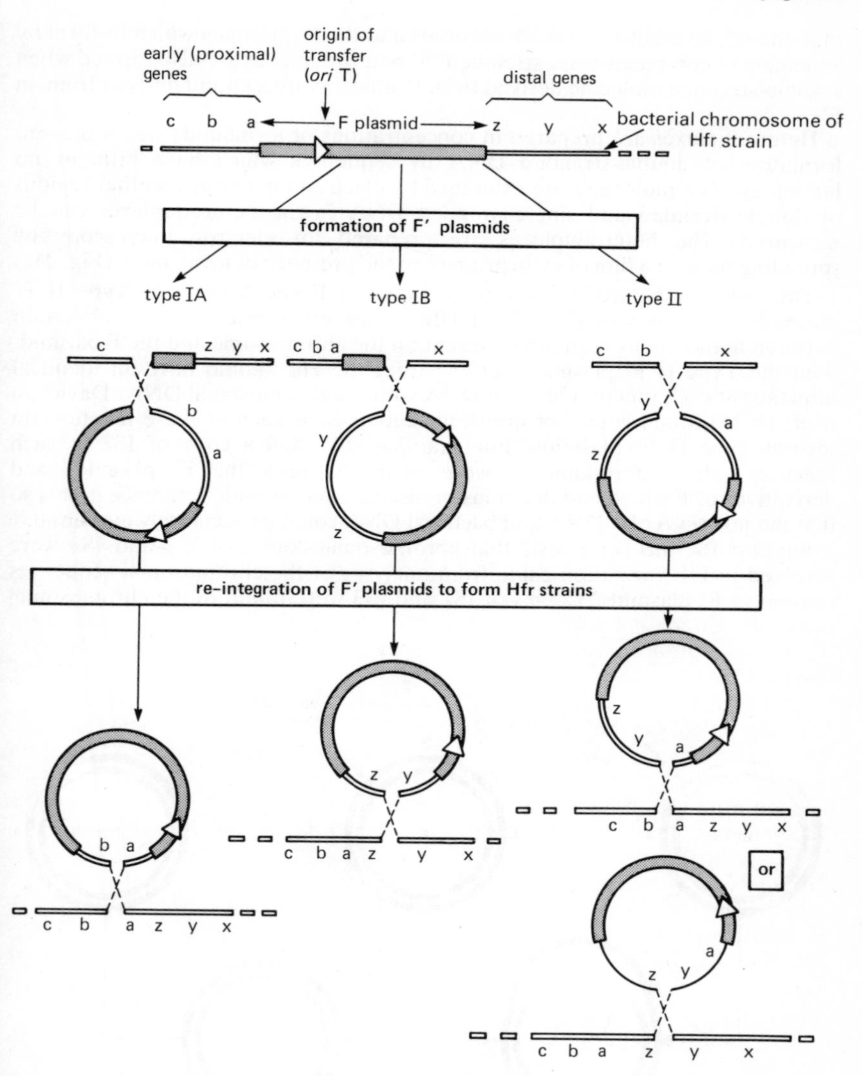

Fig. 11 Formation of F′ plasmids and formation of Hfr strains by integration of F′ plasmids into the chromosomes of secondary F′ strains. (Adapted from Low, 1972.)

preferable as these yield only two single-stranded molecules on denaturation, a linear molecule and a circular molecule. Plasmids with many nicks are unsuitable because they become fragmented into large numbers of pieces when they are denatured.) The mixture of single-stranded molecules is then allowed to reanneal. In addition to the F plasmids and the F′ plasmids which re-form by annealing of complementary strands, F/F′ heteroduplexes are also formed when a single-stranded molecule derived from F anneals with a strand derived from an F′

Heteroduplexes are prepared in concentrations of formamide which prevent formation of double-stranded DNA by sequences which have little or no homology. The molecules are examined by electron microscopy so that regions of double-stranded and single-stranded DNA in the heteroduplexes can be measured. The heteroduplexes are prepared for electron microscopy by spreading them in a film of cytochrome c in the presence of formamide (Fig. 2).

The types of heteroduplex formed between F and Type I or Type II F′ plasmids are shown in Fig. 12. If Hfr strains are formed by recombination between homologous sequences present on the chromosome and the F plasmid, then the Type II F′ plasmids derived from the Hfr should have an identical sequence at each junction of the F DNA with the chromosomal DNA. Davidson *et al.* (1975) found copies of insertion sequences at each of these junctions in several Type II F′ plasmids. For example, F13 had a copy of IS2 at each junction. The heteroduplexes were made between the F′ plasmids and derivatives of F which had deletions or insertions to provide reference points so that the junctions of F DNA and bacterial DNA could be accurately measured.

Support for the hypothesis that chromosomal copies of IS2 and IS3 were involved in Hfr formation came from analyses of the chromosomal sequences present in F′ plasmids. Copies of IS2 and IS3 were found in the chromosomal

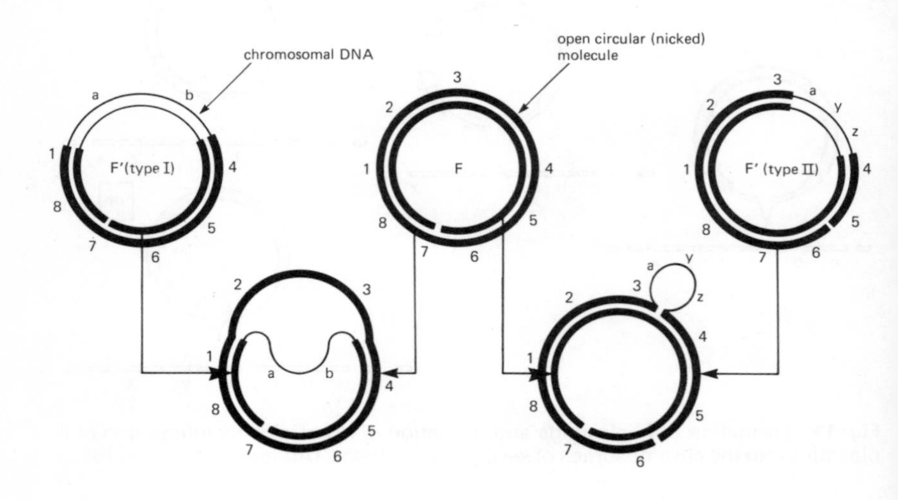

Fig. 12 Heteroduplexes formed by F and F′ plasmids

DNA at positions predicted from the sites of integration of F to form Hfr strains. Further data on the insertion sequences responsible for the formation of Hfr strains and F′ plasmids are listed in Table 3. It should be mentioned that most of the IS-elements referred to in this table have been identified from their homology with other insertion sequences. Transposition of the elements has not been demonstrated in most cases.

Hfr formation may come about simply because the insertion sequences present on the F plasmid and on the bacterial chromosome provide sites of homology which are acted upon by the *recA*$^+$-dependent recombination mechanisms of the bacterial cell. That is, Hfr formation may occur irrespective of the peculiar properties of the insertion sequences. However, Hfr formation can occur in *recA* mutants, implying that insertion sequences can themselves promote integration, a property which is presumably related to their capacity for *recA*$^+$-independent transposition. Fusion of two replicons occurs at step 5 in the model for transposition shown in Fig. 16.

Instability of F′ plasmids F′ plasmids frequently gain or lose long sequences of DNA. These changes may occur either as a result of recombination between sites within the plasmid or by integration into, and excision from, the chromosome.

Type II F′ plasmids which have a γδ sequence at each end of the F plasmid sequence frequently dissociate into two circular molecules; one circle comprising the complete sequence of the F plasmid and the other comprising the bacterial sequences with which the F plasmid was previously associated. The chromosomal sequences do not constitute a replicon and are therefore eventually lost from the culture. Dissociation is caused by recombination between the two directly

Table 3 Hfr strains and F′ derivatives

Hfr	origin (min)	Direction of transfer	Insertion sequences involved in Hfr formation		F′ derivatives†	type
			F	chromosome		
P3	12	clockwise	IS*3* ($\alpha_1\beta_1$)††	IS*3* ($\alpha_4\beta_4$)	F100, F152	I
AB133	83	clockwise	γδ	?	F14	II
P804	7	clockwise	IS*3* ($\alpha_1\beta_1$)	IS*3* ($\alpha_3\beta_3$)	F42	I
Ra-2	86	anticlockwise	γδ	?	KLF5	II
13	8	anticlockwise	IS*2*	IS*2*	F13	II
OR21	8	anticlockwise	IS*2*	IS*2*	ORF203	II
OR7	10	clockwise	γδ	?	F120	I
OR11	10	clockwise	IS*3*($\alpha_1\beta_1$)	IS*3*($\alpha_3\beta_3$)	pRH7	II

† The F′ derivatives listed are those which have been analysed by electron microscopy, as described in the text, and which have been used to determine the insertion sequences involved in forming the parental Hfr strains.

†† αβ sequences refer to IS3 elements on the F plasmid and *E.coli* chromosome (see Fig. 6). (From Ohtsubo, E., & Hsu, M-T. (1978). *Journal of Bacteriology* 134: 778–94; Deonier, R. C., Oh, G. R., & Hu, M. (1977). *Journal of Bacteriology* 129: 1129–40.); and Timmons, M. S., Bogardus, A. M., & Deonier, R. C. (1983). *Journal of Bacteriology* 153:395–407.

repeated insertion sequences which form the two boundaries of plasmid and chromosomal DNA in Type II F′ plasmids. Each of the two circles formed by recombination between directly repeated insertion sequences in Type II F′ plasmids retain one of the IS-elements. Recombination between the directly repeated γδ sequences in the F-prime plasmids F14 and KLF5 occurs frequently even in a *recA* mutant.

In the strain in which it was formed (called the *primary* F′ strain) an F′ plasmid has bacterial genes which are missing from the chromosome. The presence of the bacterial genes attached to F will not therefore create additional homology between the plasmid and the chromosome (apart from any IS-elements which may occur in the chromosomal sequences attached to the F plasmid). When the F′ is transferred to another strain of *E.coli* (to form a *secondary* F′ strain) much of the F′ will be homologous with the corresponding region of the bacterial chromsome. Recombination between duplicated chromosomal sequences leads to the frequent integration and excision of the F′ in such a strain (unless it is *recA*). However, excision of the integrated F′ may occur by recombination at sites which were not involved in the original integration so that the F′ becomes altered.

Insertion sequences responsible for F′ formation Fig. 11 shows that F′ plasmids are formed when recombination occurs at sites which were not involved in forming the parental Hfr. Many of these exicision sites are insertion sequences. Analyses of heteroduplexes showed that the Hfr strain HfrP3 was formed by recombination between two IS*3* elements ($\alpha_1\beta_1$ on the F plasmid and $\alpha_4\beta_4$ on the chromosome). F100 and F152 are Type I F′ plasmids derived from HfrP3. Both plasmids have a deletion of F DNA ending at the γδ insertion sequence, indicating that F′ formation occurred by recombination of this sequence with a chromosomal sequence. The chromosomal excision point of F120, another Type I F′ plasmid, corresponds to an IS*3* element ($\alpha_4\beta_4$). The end points of the deletion of F plasmid DNA in this F′ indicates that the deletion resulted from recombination between IS3 and an apparently unrelated F plasmid sequence. Type II F′ plasmids are formed by excision at two chromosomal sites. The two chromosomal sites involved in the formation of F′ plasmids ORF203 and pRH7 were copies of IS*3* and IS*5* respectively, but not all Type II F′ plasmids are formed by recombination between two homologous IS-elements.

Transfer of chromosomal genes by Hfr strains F plasmids cease to replicate autonomously when they are integrated in Hfr strains, but the genes of the *tra* region are still expressed. Conjugal transfer of an autonomous F plasmid begins at *oriT* (see Fig. 6). The 5′ end of a single-stranded F is the first to enter the recipient; the DNA is transferred in the order *oriT—oriV—IS2*. When the F plasmid is integrated, only part of the F plasmid sequence is initially transferred followed by the chromosomal sequences adjacent to the F plasmid in the Hfr. The length of the F sequence which is initially transferred depends on which of the four F insertion sequences was involved in forming the Hfr. None of the *tra* genes are transferred initially. In fact, they are only very rarely transferred at all by Hfr donors since the connexion between donor and recipient is usually broken before the most distal genes of the Hfr are transferred. Thus, recipients usually receive only part of the F plasmid, the region from *oriT* to one of the IS-elements.

The direction of transfer of chromosomal genes by an Hfr strain depends on the orientation of the integrated F plasmid, which in turn depends on the orientation of the insertion sequences involved in forming the Hfr. A given Hfr strain transfers chromosomal genes from a fixed point in either the clockwise or anti-clockwise direction. The use of Hfr strains for mapping the positions of chromosomal genes depends on this fact and on the rate of chromosome transfer remaining approximately constant throughout conjugation.

When Hfr strains are used to map genes, each gene is given a position according to its *time-of-entry* into the recipient. Whether a gene is transferred early or late during conjugation depends on the particular Hfr which is being used. But if the origins and directions of transfer of the Hfrs are known, then the results from different experiments can be compared. The genetic map of *E.coli* K-12 is 100 minutes, the time required for transfer of the entire *E.coli* chromosome by conjugation

Mapping a gene usually depends on being able to recognize recombinants of the recipient strain which have an altered phenotype. About 10% of transferred chromosomal fragments recombine with the recipient chromosome to form recombinants which can usually be detected as colonies on appropriate growth media. For example, transfer of the $thrA^+$ allele into a $thrA^-$ recipient can be detected using agar plates containing all the growth requirements of the recipient except for threonine; only $thrA^+$ recombinants of the recipient strain can form colonies on such a medium. (An appropriate antibiotic is also included in the medium to prevent growth of the donor.)

The time-of-entry of a gene is determined by an *interrupted mating* experiment. Donor and recipient strains are mixed together in broth and samples are taken at intervals from the culture. These samples are immediately diluted and shaken violently to separate mating bacteria and to prevent any more mating. Appropriate dilutions of the bacteria are then spread onto plates to select particular types of recombinant.

The results of an interrupted mating experiment by Jacob and Wollman, who first thought of using this technique to map genes, are shown in Fig. 13. When interpreting data from interrupted mating experiments it is assumed that recombinants which have alleles from the donor are formed only if the donated gene has been received by the time mating was interrupted. Fig 13 shows that gal^+ recombinants were not found in the culture until the bacteria had been mating for at least 25 min; none of the samples of bacteria taken before this time had received the gene for galactose utilization from the donor Hfr strain. However, genes specifying resistance to azide and to bacteriophage T1 were transferred to the recipient after about 10 min. Extrapolation of the curves to the abscissa, as shown, gives the earliest times-of-entry of the genes and their positions on the *E.coli* K-12 map.

The slopes of the curves for proximal genes in Fig. 13 are steeper than those of distal genes. This is because not all matings begin at exactly the same moment when donor and recipient are mixed together in broth, so there will be a corresponding scatter in the times-of-entry of a particular gene. The rate of DNA transfer from each donor varies. This increases the asynchrony of times-of-entry of distal genes more than proximal genes.

Fewer recombinants are obtained for distal genes than for proximal genes. In the experiment illustrated in Fig 13 about three times more azide-resistant recombinants than *gal*$^+$ recombinants were obtained. This is because donors and recipients usually separate before the entire chromosome has been transferred: it is not because distal genes are less likely to recombine with the recipient chromosome once they have been transferred.

Distal genes cannot be mapped accurately because they are transferred more asynchronously than proximal genes and few recombinants are obtained. Once the approximate position of a gene is known, an Hfr which transfers the gene proximally can be used to obtain more accurate mapping. In addition, different Hfr strains transfer DNA at slightly different rates. The times-of-entry of genes known to be close to the gene to be mapped should also be determined in

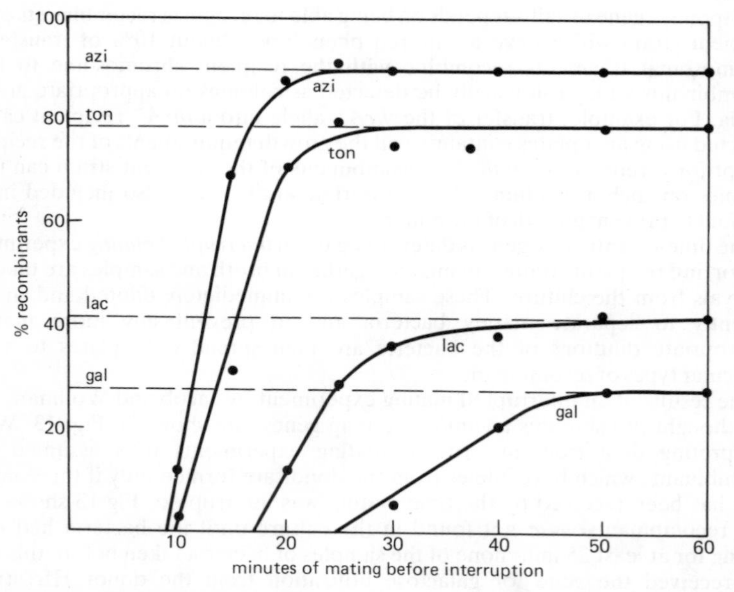

Fig. 13 Mapping chromosomal genes by interrupted mating. Donor: HfrH (*thr*$^+$ *leu*$^+$ *gal*$^+$ *lac*$^+$ *ton*^{r} *azi*^{r} *str*^{s}); recipient: F− (*thr*− *leu*− *gal*− *lac*− *ton*^{s} *azi*^{r}). The donor is resistant to bacteriophage T1 and to sodium azide. It is able to utilise lactose or galactose as carbon sources and is sensitive to streptomycin. The recipient is unable to grow on minimal salts-glucose agar unless threonine and leucine are added. Samples taken at intervals were agitated to separate mating bacteria. Prototrophic (*thr*$^+$, *leu*$^+$) recombinants of the recipient were selected on minimal salts agar containingstreptomycin to prevent growth of the donor. The recombinants were then tested for the presence of four other donor genes, *gal*$^+$, *lac*$^+$, *ton*^{r} *and azi*$^+$. (Data of Jacob & Wollman, 1961)

interrupted mating experiments so that the gene can be mapped more precisely on the *E.coli* chromosome. However, in the present state of knowledge of the *E.coli* chromosome the great value of the interrupted mating technique is that it provides a rapid estimate of the approximate position of a gene. A more precise estimate of its position in relation to closely linked genes can then be made by three-factor crosses using the generalized transducing phage. P1.

The circular chromosome of *E.coli* K-12 has a total length of 100 min, corresponding to a molecular weight of 2.7×10^9 or 4.1×10^6 base pairs. One minute on the map therefore corresponds to 41 000 base pairs.

Transfer of chromosomal genes by F′ plasmids Chromosomal genes attached to F in the form of an F′ are transferred at high frequency by the plasmid, but other chromosomal genes may also be transferred frequently because F′ plasmids readily form Hfr strains by recombination with the bacterial chromosome (Fig. 11).

When an F′ plasmid is transferred from the strain in which it was formed (primary F′ strain) to another strain (secondary F′ strain) this strain becomes diploid for those chromosomal genes occurring on the F′. Recombination occurs frequently between homologous sequences present on the F′ and the chromosome, to form Hfr strains. These Hfr strains will be related to the parental Hfr which gave rise to the F′. An Hfr formed by a Type IA F′ transfers a set of chromosomal genes as proximal markers and another set of the *same* genes as distal markers. The proximal genes are the same genes which were transferred proximally by the parental Hfr strain. The proximal genes of an Hfr formed by a Type IB F′ are the same genes which were distal in the parental Hfr. Genes which were transferred early by the parental Hfr follow the few genes which were transferred last by this strain. The time-of-entry of genes transferred by a Type IB F′ will therefore be delayed in comparison with the parental Hfr. The proximal genes transferred by an Hfr derived from a Type II F′ will be the same as those transferred proximally by the parental Hfr.

One type of chromosome transfer is a consequence of an F′ being formed rather than a means of transfer promoted by an F′ itself. When a Type I F′ is formed, part of the F plasmid remains in the *E.coli* chromosome. This forms a *sex-factor affinity* locus (*sfa* locus). If the Type I F′ which has been formed is removed from a primary F′ strain by curing, another F plasmid can be transferred into the strain. This plasmid frequently forms an Hfr at the *sfa* site because of its homology with the fragment of F which is left at this locus. The Hfr strains formed by recombination at *sfa* have the same orientation as the parental Hfr strain. (It is sometimes impossible to obtain a cured derivative of a primary F′ strain because the F′ may have chromosomal genes which are essential for cell growth.)

Transfer of non-conjugative plasmids by conjugative plasmids: plasmid mobilization

Conjugative plasmids often transfer non-conjugative plasmids at about the same frequency as they transfer themselves. For example, more than 90% of recipients of the F plasmid from a donor harbouring both F and the non-conjugative plasmid ColE1, also receive ColE1. A small proportion (about 5%) of recipients become $ColE1^+$ but remain F^-. Non-conjugative plasmids may be mobilized very efficiently by some conjugative plasmids but at frequencies of only about 10^{-6} by others. ColK-K235 is nonconjugative (molecular weight 5×10^6) and is efficiently mobilized by de-repressed I-like plasmids such as R144*drd3* (about 90% of recipients receiving the R plasmid also receive ColK), but much less efficiently by F or F-like plasmids (co-transfer frequencies of less than 10^{-4}). In contrast, ColE1 is efficiently mobilized by both F-like and I-like plasmids.

Mobilization depends neither on co-transfer of the conjugative plasmid nor on the bacterial *recA*$^+$ gene-product. ColE1 is mobilized by *traI* mutants of F which cannot transfer copies of the F plasmid.

Nonconjugative plasmids such as ColE1 and ColK which can be efficiently mobilized specify at least one product which is essential for mobilization. This protein acts in *trans* so that ColK can complement mobilization-defective mutants of ColE1, enabling the ColE1 mutants to be transferred efficiently by I-like plasmids.

Clewell and Helinski found that many plasmids, including ColE1, can be isolated in the form of a *relaxation complex* in which supercoiled plasmid DNA is associated with proteins which can act as endonucleases. When relaxation complexes are treated with proteases or alkali, for example, the proteins of the complex are activated so that they break one of the polynucleotide strands of the plasmid which then assumes the open-circular (*relaxed*) form. Many plasmid isolation methods activate relaxation proteins, reducing the yield of supercoiled plasmid DNA. The nick produced by the relaxation proteins is made at a specific site in ColE1; it is always the *heavy* strand which is nicked and the break is about 300 nucleotides from the origin of ColE1 replication. Plasmid mutants which have altered relaxation proteins cannot be mobilized. Upon relaxation, one of the relaxation proteins remains attached to the 5′ end of the nicked strand. It has been suggested that it acts as a pilot protein, leading the 5′ end of the plasmid into the recipient.

Non-conjugative plasmids which are not efficiently mobilized by F can nevertheless be mobilized at very low frequencies (about 10^{-6} per donor). Plasmids which have been mobilized by F at these low frequencies acquire an insertion of the $\gamma\delta$ IS-element. During the process of transposition of $\gamma\delta$ from F to the nonconjugative plasmid, it appears that the two plasmids become fused together (see step 5 in Fig. 16) so that the nonconjugative plasmid is transferred along with F. The transposition process is presumably completed in the recipient; the nonconjugative plasmid, now having an insert of $\gamma\delta$, becomes separated from F.

Chromosome transfer by F^+ cultures of *E.coli* K-12

Strains containing autonomous F plasmids can also transfer chromosomal genes, albeit at much lower frequencies than Hfr donors. All chromosomal genes are transferred by F^+ strains at a frequency at about 10^{-6} per donor. Although this mode of chromosome transfer was the first to be described (the strains used by Tatum and Lederberg were F^+), it is the least well understood.

Following the discovery of Hfr strains, it was suggested that chromosome transfer by F^+ cultures was due to the few Hfr cells which happen to exist in the population. But it was found that Hfrs were responsible for only about 20% of recombinants produced by F^+ donor cultures. Recombinants produced by F^+ donor cultures were almost invariably F^+; they had received a complete F plasmid as well as the chromosomal genes. This also appeared to be inconsistent with Hfr cells being responsible for chromosome transfer, as most recombinants produced by Hfr donors are F^-. This could be explained if F^+ donors transferred copies of F into recipients which had received chromosomal DNA from the new Hfr cells in the culture. However, experiments with mixtures of Hfr, F^+ and F^- cells rule out this possibility.

Conjugation and chromosome transfer in other bacteria

Genetic analysis of the chromosome of any bacterial species is made much easier if a conjugative plasmid can be used to transfer genes between mutant strains. Transformation and transduction can also be used to map genes in some genera; many genes have been mapped on the *Bacillus subtilis* chromosome by transformation. However, conjugative plasmids provide the simplest and most useful means of transferring chromosomal genes from one strain to another.

The F-plasmid of *E.coli* K-12 can be used to map genes in several members of the *Enterobacteriaceae*. F integrates into the chromosomes of several *Salmonella* species to form Hfr strains. At least 430 genes have been mapped on the *S. typhimurium* chromosome by interrupted mating experiments and by transduction, mainly by phage P22. F can also form Hfr strains and promote chromosome transfer in other members of the *Enterobacteriaceae* such as *Citrobacter freundii* and *Erwinia chrysanthemi,* a plant pathogen.

F cannot however, be stably maintained in all members of the *Enterobacteriaceae,* so other plasmids must be used to transfer genes in these bacteria and in nonenterics. Plasmids of the IncP1 group, which were originally found in strains of *Pseudomonas aeruginosa,* can not only transfer genes between pseudomonads, but also between many other genera of Gram-negative bacteria. IncP1 group plasmids have a particularly wide host-range and can be maintained in most Gram-negative bacteria. They are sometimes called *promiscuous plasmids* for this reason. RP4 and R68.45 are the P group plasmids which have been most frequently used to transfer chromosomal genes between Gram-negative bacteria. They both code for resistance to carbenicillin (and other penicillin derivatives), kanamycin and tetracycline.

RP4 and R68.45 have been used to map the positions of numerous genes of bacteria such as *Acinetobacter calcoaceticus, Rhizobium leguminosarum and Rhizobium meliloti,* and to demonstrate that these bacteria have circular chromosomes. They usually transfer the chromosome from several origins and do not appear to form stable Hfr strains. However, by creating a region of homology between the plasmid and the bacterial chromosome, the plasmid can be made to behave like an F′ and to transfer the chromosome mainly from a single origin. Three procedures have been used to make a region of homology in P group plasmids.

DNA from the bacterial chromosome can be added to the plasmid by *in vitro* recombination using restriction endonucleases and DNA ligase. In *E.coli*, RP4 containing a piece of *E.coli* chromosome behaves like an F′ and transfers the chromosome from an origin corresponding to the region of homology between plasmid and chromosome. Mobilization of the *E.coli* chromosome by RP4-prime plasmids depends on the *recA*$^+$ gene of the bacterium.

A second method for creating homology between plasmid and chromosome is to add a copy of phage Mu chromosome to each replicon. Mu itself is a kind of transposon; it can become inserted into many unrelated DNA sequences. An RP4-Mu plasmid behaves like an F′ and transfers chromosomal DNA from the site of Mu insertion in the chromosomes of *E.coli* or *Klebsiella pneumoniae*. A third procedure for creating homology is to insert a transposon into both plasmid and chromosome to bring about enhanced chromosome transfer from the site of transposon insertion.

The usefulness of P group plasmids has also been increased by the isolation of drivatives which transfer chromosomal genes at higher frequencies. One such plasmid is R68.45 which was derived from R68. R68.45 has a duplication of a 2.1kb sequence present on R68. The new sequence on R68.45 is an insertion sequence (IS*21*), although the single copy present on R68 is inactive in transposition. It is thought that R68.45 transfers chromosomal genes at a high rate because IS*21* forms a cointegrate (see Fig. 16) during transposition. R68.45 does not form stable Hfr strains, but R68.45-primes comprising chromosomal genes from *Pseudomonas* and *Rhizobium* can be made, indicating that the plasmid can integrate into the chromosome.

Many aspects of chromosome transfer in *Pseudomonas aeruginosa* have been investigated by Holloway. Several plasmids were found to transfer the *Pseudomonas aeruginosa* chromosome from a unique origin. The strains behaved like Hfrs although there was no evidence for stable integration of the plasmid and it is not known whether a homologous sequence, or perhaps an insertion sequence, exists at the origin. Orientated chromosome transfer promoted by FP2 and other plasmids in *Pseudomonas aeruginosa* has been used to map numerous genes and to demonstrate that the chromosome is circular.

Conjugative plasmids have been found in several genera of Gram-positive bacteria, including *Bacillus, Streptococcus, Streptomyces* and *Clostridium*.

The conjugal transfer mechanism of streptococcal plasmids appears to be fundamentally different to that of plasmids such as F found in Gram-negative bacteria. Two groups of streptococcal plasmids can be distinguished (see Clewell *et al.* 1984). R plasmids such as pAMβ1 have a molecular weight of about 17×10^6 and transfer most efficiently when mated on a solid surface such as a membrane filter or an agar plate. Some of these plasmids can transfer to other Gram-positive species such as *Lactobacillus, Staphylococcus* and *Bacillus*. A second group of plasmids, which includes pAD1 (molecular weight 35×10^6) for example, have been isolated only from *Streptococcus faecalis*. These transfer well in liquid media. Their peculiarity is that cells harbouring them respond to mating signals (sex pheromones) released from recipient bacteria (certain streptococci and staphylococci). The sex pheromones are small peptides (two which have been sequences have eight amino acids) which induce the donor to produce a protein 'adhesin' which sticks the donor and recipient cells together. Transfer of DNA takes place within these 'mating aggregates'. Once the plasmid is transferred to the recipient, the cells stop making the particular pheromone which induces transfer of the plasmid they have received. The plasmids encode a mechanism for chemical modification of their corresponding pheromone so that the active form is no longer released. It has been suggested that the pheromoues originally had another function (and perhaps still have); the plasmids have simply evolved a mechanism for using their presence as an indicator of potential recipient cells.

Several conjugative plasmids found in *Streptomyces* are also transferred at high frequencies. Two conjugative plasmids found in *Streptomyces coelicolor,* SCP1 and SCP2, have been investigated by Hopwood. SCP1 specifies the synthesis of an antibiotic, methylenomycin. It can become integrated into the bacterial chromosome to form an Hfr and in this form can transfer chromosomal genes at high frequencies (in optimal condition, almost 100% of recipients receive donor chromosomal DNA). SCP1 has been used to map the positions of numerous genes on the *Streptomyces coelicolor* chromosome and to demonstrate its circularity. SCP1 can also acquire chromosomal DNA to form SCP1-prime plasmids. Their existence can be demonstrated by genetic techniques, but complete molecules of SCP1-prime or SCP1 have not been isolated; SCP1 appears to have a molecular weight of about 200×10^6. SCP2 is a low molecular plasmid (MW = about 20×10^6). It brings about genetic recombination at low frequencies (about 10^{-7} per donor). A variant called SCP2* causes a higher frequency of recombination. Only a small region (3000–10,000 base pairs) of streptococcal plasmids such as SCP2 is needed for conjugation, compared with the 35,000 base pairs needed to encode all the proteins needed for F transfer.

Summary

The first property of a plasmid to be recognised was the ability of the F (Fertility) plasmid to transfer chromosomal genes between strains of *E.coli.* The transfer and recombination of genes in *E.coli* K-12 was discovered by Tatum and Lederberg in 1946. It was subsequently found that the F plasmid could also transfer itself to strains of *E.coli.* The products of at least 23 *tra* genes on the F plasmid are involved in conjugation. Almost all these genes are included within a single *tra* operon. Transcription of the *tra* operon of most plasmids which are closely related to F (F-like plasmids) is subject to negative control by repressors, but expression of the *tra* operon of F is not repressed. F does not specify an effective repressor although it retains an operator site at which the repressors of F-like plasmids can act to cause *fertility inhibition.*

The products of several *tra* genes are required to produce a protein tube called a sex pilus. Each bacterium containing an F plasmid produces one or two sex pili which can be up to 20 μm long. Although the sex pilus is essential for transferring DNA by conjugation, it is not clear whether it acts as a long thin tube through which DNA passes from donor to recipient, or whether its primary function is to contract once contact has been made with a recipient in order to bring the two cell walls close together. During conjugation, a single-stranded molecule of F DNA is transferred into the recipient, in which the complementary strand is synthesized.

Chromosomal genes are transferred at high frequency from cells in which the F plasmid has become integrated into the bacterial chromosome to form an Hfr (*H*igh *fr*equency) strain. When an Hfr strain is used as a donor, part of the F plasmid is transferred into the recipient, followed by chromosomal DNA which is covalently attached to it. Occasionally, a copy of the entire *E.coli* chromosome is transferred (in single stranded form), but contact between donor and recipient is usually broken before this happens. Under ideal conditions, transfer of the whole chromosome takes about 100 minutes. Because a given Hfr strain always transfers chromosomal genes in the same order and at about the same rate, such strains have been extensively used to map the positions of genes on the *E.coli* chromosome according to their time-of-entry into the recipient.

Integration of F to form an Hfr takes place by recombination between an insertion sequence (IS*2*, IS*3* or γδ) on F and, in some cases at least, a homologous insertion sequence on the chromosome.

F′ (F-prime) plasmids are formed when excision of F from the chromosome of an Hfr strain involves recombination at chromosomal sites so that a circular molecule is produced in which chromosomal sequences are attached to F. The excision sites involved in forming F′ plasmids are often insertion sequences.

Many other plasmids, in addition to the F plasmid, are able to transfer copies of themselves and chromosomal genes between strains of bacteria. Some plasmids found in *Pseudomonas,* such as RP4, are able to transfer themselves to many different genera of Gram-negative bacteria and can be used to map chromosomal genes in them.

Conjugative plasmids have been found in several genera of Gram-positive bacteria: *Bacillus, Streptomyces, Streptococcus* and *Clostridium.*

References

BACHMANN, B. J. (1983). Linkage map of *Escherichia coli* K-12, edition 7, *Microbiological Reviews,* 47:180–230.

DAVIDSON, N., DEONIER, R. C., HU, and S. OHTSUBO, E. (1975). Electron microscope heteroduplex studies of sequence relations among some plasmids of *Escherichia coli.* In: *Microbiology-1974* pp. 56–65. Edited by D. Schlessinger. American Society for Microbiology, Washington.

GRINDLEY, N. D. F. and REED, R. R. (1985). Transpositional recombination in prokaryotes. *Annual Review of Biochemistry* 54:863–896.

HAYES, W. (1968). *The Genetics of Bacteria and their Viruses,* 2nd edition. Blackwell, Oxford.

JACOB, F. & WOLLMAN, E. (1961). *Sexuality and the Genetics of Bacteria.* Academic Press, New York.

LOW, K. B. (1972) *Escherichia coli* K-12 F-prime factors, old and new. *Bacteriological Reviews* 36: 587–607.

MEYNELL, G. G. (1972). *Bacterial Plasmids,* Macmillan, London.

WILLETS, N. and WILKINS, B. (1984). Processing of plasmid DNA during bacterial conjugation. *Microbiological Reviews* 48: 24–41.

4 R plasmids

Conjugative R plasmids were discovered in Japan in 1957. Until about 1950, bacillary dysentery in Japan was treated with sulphonamide, although this drug was rapidly becoming less effective because an increasing proportion of *Shigella* strains were sulphonamide-resistant. Tetracycline, streptomycin and chloramphenicol were also used to treat dysentery from about 1950, but by 1957 a small proportion (about 2%) of shigellae were found to be resistant to one or more of these antibiotics. The proportion of resistant strains increased to 13% by 1960; 9% of the strains isolated in this year were resistant to sulphonamide, chloramphenicol, streptomycin and tetracycline. Eventually, the multiply resistant strains became the predominant type of *Shigella* in Japan.

The first indication that the multiple resistance might be plasmid-specified came from analyses of the strains of *Shigella* responsible for epidemics. Both multiply resistant and fully sensitive forms of the same strain were sometimes isolated during an epidemic. Indeed, both forms were sometimes isolated from individual patients. Multiply resistant strains of *E.coli* were sometimes isolated from patients excreting resistant shigellae, so Akiba and Ochiai suggested that resistant *E.coli* might transfer DNA coding for drug-resistance to *Shigella*.

The genes for drug-resistance were shown to be transferred by a process that required cell contact; that is, they were transferred by conjugation. This process did not depend on the presence of the F plasmid although the genes specifying drug-resistance could be cured by treatment with acridine orange and were not linked to chromosomal genes. Transfer could also occur in the alimentary tract.

Following the Japanese discovery of R plasmids in enterobacteria, they were soon found in strains isolated from many parts of the world. They occur in both Gram-positive and Gram-negative bacteria and have been found in almost all the species which are pathogenic for man or animals. R plasmids are particularly common in the enterobacteria, such as *E.coli, Salmonella* and *Shigella,* and in staphylococci.

Almost all antibacterial drugs used in medicine have a corresponding resistance gene on at least one type of R plasmid (see Table 5). Drug resistance can also be specified by chromosomal genes, but the mechanism of resistance is usually different to that specified by plasmids. Only chromosomal resistance to nalidixic acid and nitrofurans has been found so far.

Two important features of R plasmids which have contributed to their rapid evolution and dispersal are their conjugative ability and the presence within their sequences of genetic elements known as *transposons*. These two aspects of R plasmids are considered first before we examine the various uses of antibiotics which have provided the selection pressure for R plasmid evolution.

Transfer of R plasmids by conjugation

Conjugative plasmids belonging to the IncP1 group which can replicate and maintain themselves in almost all species of Gram-negative bacteria may be especially important in disseminating drug-resistance to a wide range of organisms.

Many of the conjugative R plasmids found in enterobacteria belong to either the F-like or I-like classes, that is, they specify pili resembling those encoded by F or by ColI, the prototypes of the two classes. These plasmids usually have molecular weights of at least 60×10^6. R plasmids specifying other types of sex pili have been described (p.33). The smallest conjugative R plasmids so far found in the enterobacteria belong to the W incompatibility group and have molecular weights of only 20 to 25×10^6.

R plasmid mobilization Many R plasmids are not conjugative themselves but are nevertheless spread much more rapidly between strains of bacteria than are chromosomal genes. This is because they are *mobilized* (p.50), often very efficiently, by conjugative plasmids. About 50% of enterobacteria harbour conjugative plasmids which can mobilize nonconjugative R plasmids. Mobilization is also an important means of plasmid transfer in other types of bacteria. For example, about 7% of *Neisseria gonorrhoeae* strains have a conjugative plasmid which can mobilize small nonconjugative R plasmids if these happen to be present.

The transmissible drug-resistance of a strain is sometimes the result of its harbouring three or more R plasmids, some conjugative and others not. The existence of the several different R plasmids in such a strain may become apparent only after careful genetic analysis. The strain of *Salmonella typhimurium* type 29 which was responsible for the outbreak of salmonellosis described on p.74 became resistant to sulphonamide, streptomycin, ampicillin and tetracycline. It harboured at least three plasmids. Although all the resistance determinants could be transferred from *Salmonella* by conjugation, none was present on a conjugative plasmid. Their transfer by conjugation was promoted by an I-like conjugative plasmid called Δ. The genes for resistance to sulphonamide and to streptomycin were both on one plasmid (molecular weight 5.6×10^6). Ampicillin-resistance was specified by another small plasmid of about the same molecular weight. These plasmids were transferred at a high frequency from *S.typhimurium* donors because they were efficiently mobilized by Δ. Tetracycline-resistance was transferred at a much lower frequency. A strain receiving tetracycline-resistance was found to have a plasmid which was larger than Δ (67×10^6 compared with 61×10^6 for Δ). It is not clear whether the increment of 6×10^6 represents the recombination of a small plasmid with Δ or the addition of a sequence of chromosomal DNA.

Relationships between F and F-like plasmids The relationships between these plasmids, which were first indicated by *fertility inhibition* (p 28), are also evident from analyses of DNA molecules using the heteroduplex method. The F-like plasmids R1, R100 (also called NR1 or R222), R6 (R6–5 is a tetracycline-sensitive derivative of R6) and ColV, I-K94 form heteroduplexes with F indicating that much of their DNA is similar in sequence to that of the F plasmid. R1, R100 and

R6 all belong to incompatibility group FII. The F plasmid and ColV, I-K94 belong to incompatibility group FI.

Diagrams of the heteroduplexes formed when a single-stranded DNA molecule of F is annealed with a single-stranded molecule derived from R1 or R6–5 are shown in Fig. 14. The positions of the genes specifying conjugation and antibiotic-resistance are indicated. Some of the main conclusions which can be drawn about the heteroduplexes are the following:

1 In agreement with the genetic evidence, F-like plasmids and F are homologous over most of the region which specifies proteins required for conjugation. The use of the term *homology* in connexion with heteroduplexes does not imply that the sequences which anneal to form double-stranded DNA are exactly complementary; some mismatching of bases may occur in the double-stranded regions.

2 The loops of single-stranded DNA which emerge from double-stranded homologous regions indicate that relatively large rearrangements of DNA sequences have occurred during the evolution of these plasmids from common

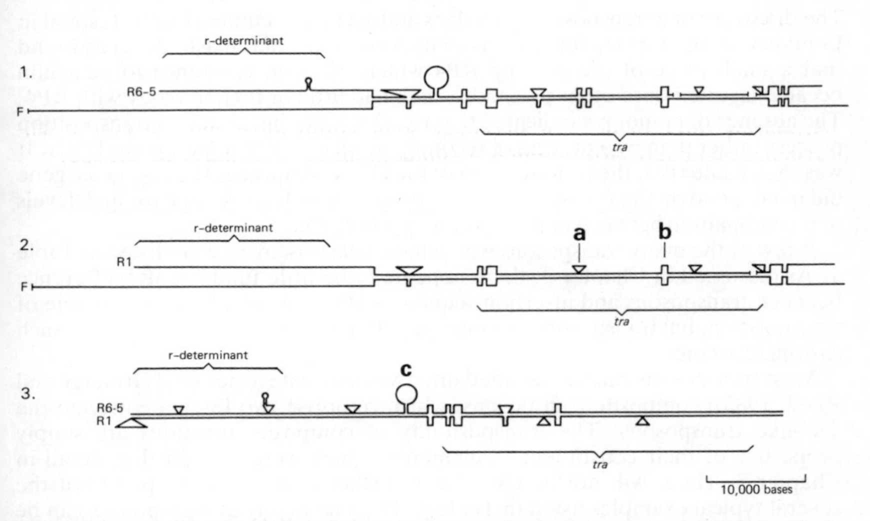

Fig. 14 Heteroduplexes formed between related F-like plasmids: 1, R6–5 and F; 2, R1 and F;3, R6–5 and R1. Regions of homology between the plasmids are indicated by close parallel lines. The type of single-stranded DNA loop shown at (a) is formed when a DNA fragment has been either deleted or added to one of the plasmids. The type of loop shown at (b) indicates that an inversion or substitution of DNA has occurred at this point during the evolution of the plasmids from a common ancestor. (c) A stem-loop structure indicates an inverted repeat (as occurs in Tn *5*, for example). Genes coding for resistance to antibiotics map within the region labelled 'r-determinant'. Genes for transfer by conjugation map within the region labelled '*tra*'. (Redrawn from Sharp, P. A., Cohen, S. N. & Davidson, N. (1973). *Journal of Molecular Biology* 73: 235–55)

ancestors. The type of single-stranded loop shown at *a* in Fig. 14 indicates that an addition or deletion of several hundred base pairs has occurred. The region of nonhomology shown at *b* indicates that an inversion or substitution of DNA has occurred.

Tetracycline-resistance is conferred by the transposon Tn*10* in R6. R6–5 does not confer tetracycline-resistance because a copy of the IS*10*-element, which occurs at each end of Tn*10*, has become inserted into one of the transposon genes needed for tetracycline-resistance.

3 With the exception of the tetracycline-resistance determinant, the resistance genes are clustered in one region, the 'r-determinant'. This cluster of genes is flanked by copies of the insertion sequence IS1 in each of the plasmids.

Plasmids R1, R100 and R6 are closely related to each other even though they were found in strains isolated in different countries. R1 was isolated in London, R100 in Japan and R6 in Austria.

Transposons

The discovery of a transposon by Hedges and Jacob at Hammersmith Hospital in London was a major step forward in our understanding of R plasmids. They found that a small piece of the plasmid RP4 which specified resistance to penicillin became inserted onto other plasmids which had little or no homology with RP4. The absence of homology indicated that the DNA was inserted by a transposition process rather than by a process of recombination between homologous DNA. It was shown later that the transposition of the DNA comprising the resistance gene did not depend on the *E.coli recA*$^+$ gene-product which is essential for high levels of recombination between homologous DNA in *E.coli*.

A few of the many transposons which have been discovered are listed in Table 4. As discussed in Chapter 3, there appears to be little fundamental difference between transposons and insertion sequences (IS-element); both are capable of transposition, but transposons, in contrast to IS-elements, also encode a trait such as drug resistance.

Most transposons can be classified into two main categories (see Grindley and Reed, 1985): composite transposons, which comprise two IS-elements, and the Tn*3*-like transposons. The transposibility of composite functions are simply properties of their constituent IS-elements, which were described in detail in Chapter 3. These will not be considered further here except to point out the several typical examples listed in Table 4. The Tn*3* family of transposons can be divided into two groups, represented by Tn*3* itself and by Tn*501*. Both groups have a similar structure, but the position of some of the elements within the transposons are different. Bacteriophage Mu, a third category of transposable element is not considered here.

Structure of Tn3 Tn*3* is one of the most extensively studied transposons. It was found on RI and codes for a β-lactamase (penicillinase) of the TEM-I type (see p.65). Transposons which are apparently identical to Tn*3* have been found on many different plasmids in a wide range of Gram-negative bacterial genera. Related transposons have also been found in Gram-positive bacteria. Tn*3*

Table 4 Transposons

Transposon	Phenotype	Size (base pairs)	Terminal repetition (base pairs)	Flanking duplication (base pairs)
Tn3-like transposons				
Tn*3*	Ap	4957	38/38	5
Tn*501*	Hg	~8200	35/38+	5
Tn*21*	Sm Su Hg	~19 500	35/38	5
IS*1721*	Tc	11 400		
Composite transposons				
Tn*5*	Km	~5700	(8/9) IS*50(IR)*	9
Tn*9*	Cm	~2500	(18/23) IS*1*(DR)	9
Tn*10*	Tc	~9300	(17/22) IS*10*(IR)	9
Tn*903*	Km	~3100	(18/18) IS*903*(IR)	9
Tn*1681*	stable-toxin	2061	(18/23) IS*1*(IR)	9
Others				
TN*7*	Tp Sm	~14 000		5

Examples of transpoons found on plasmids in Gram-negative bacteria. Key to resistance determinants: Ap, ampicillin; Sm, streptomycin; Su, sulphonamide; Km, kanamycin; Tp, trimethoprin; Cm, chloramphenicol; Hg, mercuric ion; stable-toxin, heat stable toxin found on a plasmid in *E.coli.* IR, inverted repeat; DR, direct repeat.

+ 35 out of 38 base pairs are homologous in the inverted repeat sequences.

comprises 4957 base pairs; the 38 base pairs at each end are inverted repeats.

Tn*3* codes for 3 proteins (Fig. 15). The largest of these, called the *transposase* (about 120,000 daltons) is coded for by the *tnpA* gene; mutations in this gene prevent transposition. Transposition-defective *tnpA* mutants are recessive and can be complemented by *tnpA*+ transposons. The *tnpR* gene encodes a *resolvase,* an enzyme which brings about site-specific recombination. It acts at a site located between the *tnpA* and the *tnpR* genes, a region which also has the promoters for these two genes (Fig. 15). This is called the cointegrate resolution, or *res* site. The *tnpR* protein also acts as a repressor. Transposons which have mutations in the *tnpR* gene transpose at a greater frequency than *tnpR*+ transposons. The *tnpR* protein represses synthesis of the *tnpA* product and it also represses its own synthesis. Mutant Tn*3* transposons defective in *tnpR* can be complemented in *trans* by *tnpR*+ transposons. Although *tnpR* mutants transpose more frequently, the end-products are abnormal and consist of cointegrate structures, as shown in Fig. 16 (step 5). The transposition of transposons having deletions extending into the *res* site is also blocked at step 5.

It appears that the transposase is responsible for the formation of the cointegrate which is then resolved by the *tnpR* gene-product (resolvase) into two

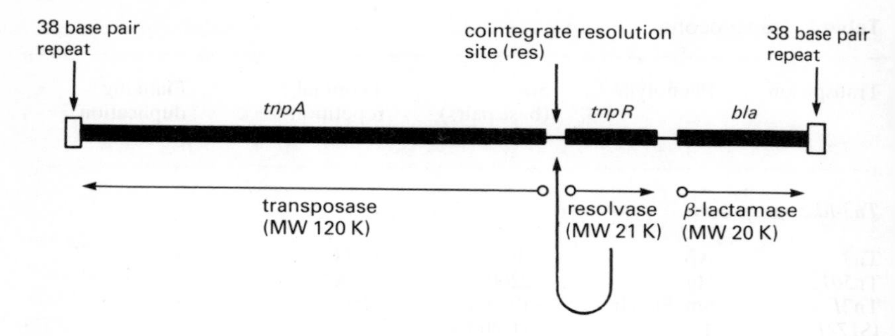

Fig. 15 Structure of the transposon Tn*3*. The arrows underneath the genes indicate the direction of transcription. The repressor coded for by the *tnpR*. (From Gill, R. E., Heffron, F. & Falkow, S. (1979). *Nature, London* 282: 797–801; Chou, J., Lemaux, P. G., Casabadan, M. J. & Cohen, S. N. (1979). *Nature, London* 282. 801–6; Heffron, F., McCarthy, B. J., Ohtsubo, H., & Ohtsubo, E. (1979). *Cell* 18: 1153–63)

separate replicons. The transposase binds to the 38 base pair inverted repeats. The *res* site has three separate regions which bind resolvase, only one of which includes the site where recombination actually takes place. The resolvase of the γδ-transposable element, which is very similar to the Tn*3* resolvase, can be split into two parts by a protease. A C-terminal fragment of 45 amino acids was found to have the specific *res* binding capacity, whereas the 140 amino acid N-terminal had the enzymatic recombination activity. The resolvase acts *in vitro* in a suitable buffer with only Mg^{2+} as cofactor. It nicks, exchanges then rejoins the DNA strands at the *res* site.

Clustering of resistance determinants Fig 14 shows that most of the resistance determinants of F-like R plasmids are clustered together. The largest group of resistance genes occurs in R1, as shown in Fig. 17. Clustering of resistance genes is common in many R plasmids, although there are exceptions; the resistance genes of RP4 are not closely linked.

If it is assumed that multiple drug-resistance usually arises through the accumulation of individual transposons on a plasmid, then clustering might occur if resident transposons were preferred sites for the addition of other transposons. Some experimental evidence supports the view that transpositions occurs more frequently near other transposons or insertion sequences; the cluster of resistance genes in the R1 plasmid has been explained in this way. R1 has two transposons, Tn*3* and Tn*4* (Fig. 18).

Clustering of resistance genes may also occur because insertion of transposons into most regions of a plasmid may inactivate an important gene. A plasmid which has an insertion of a transposon into its *tra* operon is presumably at a selective disadvantage compared with a *tra*$^+$ plasmid which has a transposon insertion in a less important gene. F-like plasmids usually have molecular weights of about 60×10^6 and the F *tra* region has a molecular weight of about 20×10^6.

Conjugative plasmids of the W incompatibility group have molecular weights of only 20 to 25 $\times$ 10^6, so insertion of transposons at all but a few sites in these plasmids may impair conjugative ability.

Whatever the reasons for clustering of resistance genes in individual plasmids, this arrangement implies that selection for one of the genes provides a strong selection for all closely-linked resistance genes. Closely-linked genes are more likely to recombine *en bloc*. Clustering of resistance genes may be a factor in the survival of a resistance gene in the absence of selection pressure (provided by the appropriate antibiotic). In addition, the close linkage of genes may increase the likelihood of forming new transposons which code for resistance to more than one antibiotic. A transposon of this type exists within the cluster of resistance genes in R1, and Tn*7* is another transposon specifying resistance to two completely different drugs, trimethoprim and streptomycin.

Instability and rearrangements of IncFII R plasmids

Although IncFII R plasmids (R1, R6–5 and R100 have been studied in most detail) can be transferred to several enterobacterial genera such as *E.coli Shigella, Salmonella* and *Proteus,* they are not equally stable in all of them. The host strain can affect the copy number of a plasmid, its rate of spontaneous loss from cells and its molecular structure. The rearrangements of IncFII R plasmids which occur in *Proteus mirabilis* depend on the presence of two directly repeated copies of the IS1 element.

When isolated from *E.coli,* almost all R100 molecules are of one type, circles with molecular weights of 58 $\times$ 10^6. The copy number of the plasmid is one to three. In *Proteus mirabilis,* the copy number of R100 increases to about five. Most of the plasmids have a molecular weight of 58 $\times$ 10^6. But as the culture enters stationary phase other forms of the plasmid predominate. One of these has a molecular weight of 45 $\times$ 10^6 and is sometimes called the RTF (*resistance transfer factor*) as it comprises all the genes necessary for conjugation. The fragment of R100 which is missing from the RTF is called the *r-determinant.* It comprises all the genes conferring antibiotic-resistance except the tetracycline-resistance determinant which is on the RTF. On entering stationary phase, the proportion of the r-determinant increases in relation to the RTF. There may be ten or more copies of the r-determinant for each copy of the RTF.

The increase in the r-determinants can be readily seen in caesium chloride density gradients because they have a higher percentage of G + C (58%) than the RTF (52%). *Proteus mirabilis* chromosomal DNA has a G + C content of 40% so it is well separated from both types of plasmid molecule in density gradients. A similar increase in r-determinants is seen when cultures are grown in the presence of chloramphenicol (there is a chloramphenicol-resistance gene on the plasmid). Presumably, cells which have higher numbers of r-determinants are selected because they are more resistant to chloramphenicol.

Many of the R plasmids in stationary phase cultures of *Proteus,* or in cultures

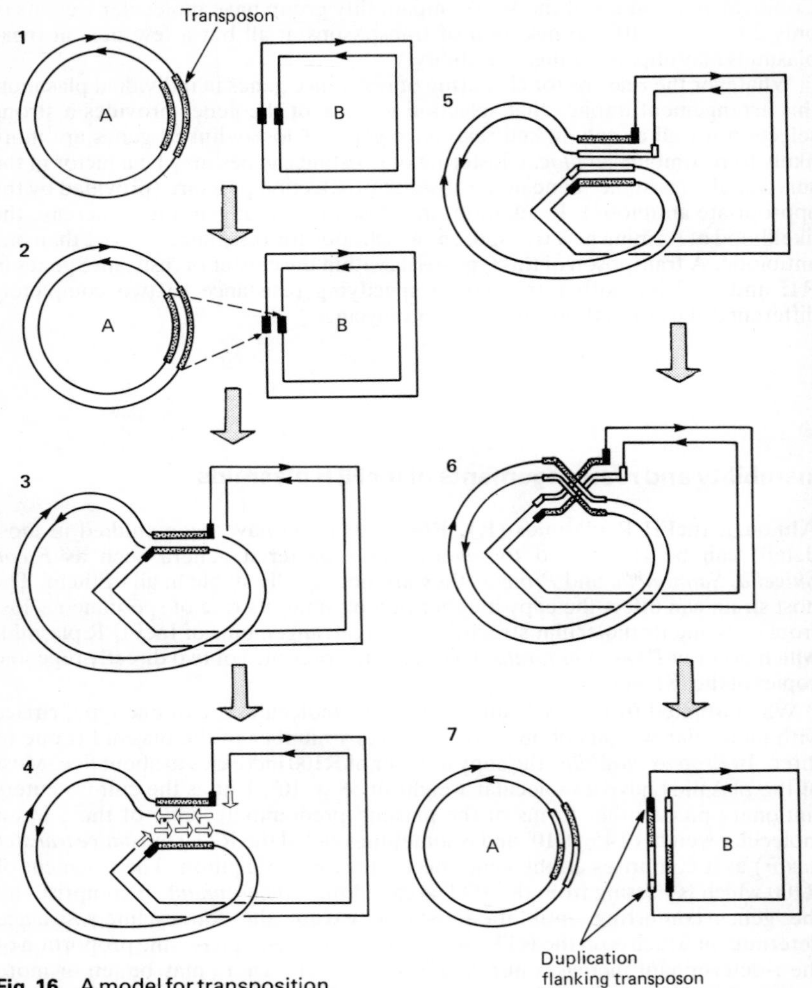

Fig. 16 A model for transposition

1 A is the donor of the transposon and B is the recipient molecule.

2 Cleavages occur at the 3′ OH ends of the transposon on A and staggered nicks, producing ends with 5′ phosphate groups, are made in B. (The nicks in B would be five base pairs apart in the case of Tn*3*.) Both these reactions probably involve the transposase coded for by the transposon. Donor strands ending in 3′ OH groups are ligated to the 5′ phosphate ends of the recipient strands.

3 Hydrogen-bonding between the strands of the transposon holds the molecule together.

4 DNA synthesis begins at the 3′ OH end(s) of B, replicating the short sequence of B which is to become the flanking duplication and the transposon.

5 A ligation reaction at the end of each new strand links A and B together, forming a *cointegrate*.

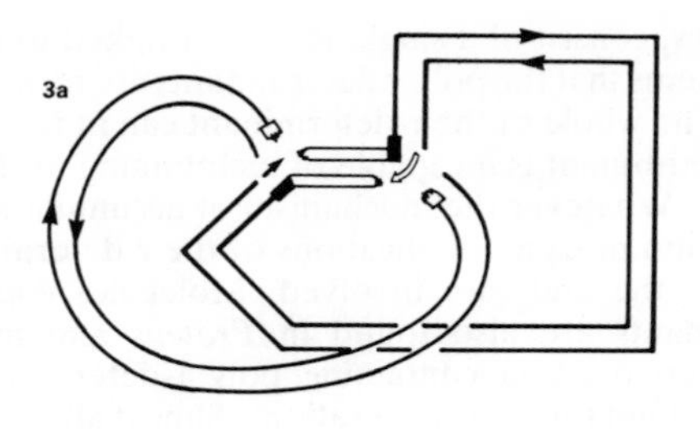

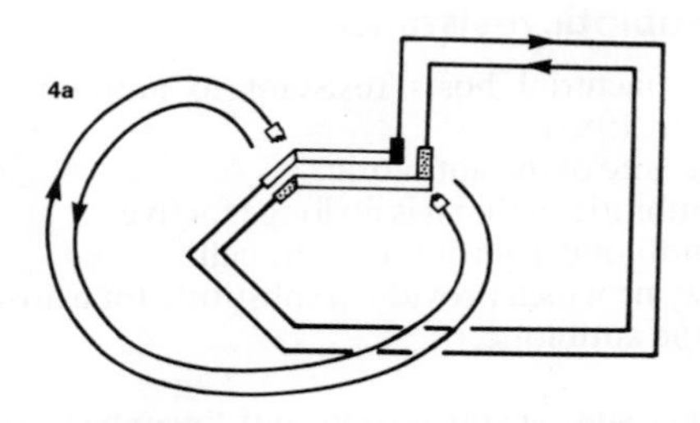

Fig. 17 A model for simple transposition without duplication. 3a. The structure shown in Fig. 16 (step 3) can give rise to a simple insertion as follows. DNA synthesis begins at the 3′OH ends of B, the single-stranded DNA attaching the transposon to the donor (A) is broken and the broken end is joined to transposon B, to form the structure shown as 4a.

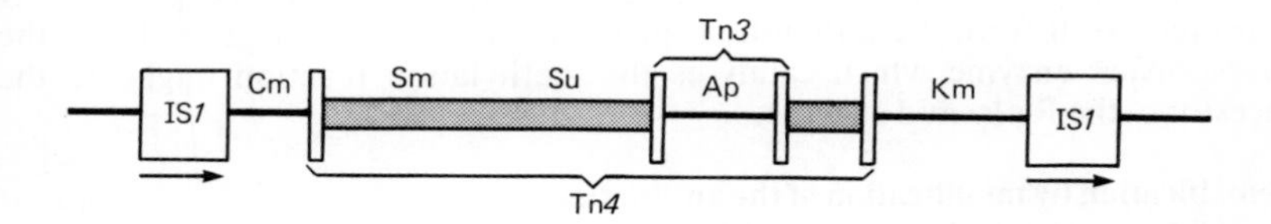

Fig. 18 Positions of genes coding for drug-resistance within the r-determinant region of R1 *drd*19 (a derivative of R1 which is de-repressed for transfer). The region contains two transposons, Tn*3* and Tn*4*. The resistance determinants are: Cm, chloramphenicol; Sm, streptomycin; Su, sulphonamide; Ap, ampicillin; Km, kanamycin. The r-determinant region has a copy of the insertion sequence IS1 at each end. Both IS-elements are in the same orientation. (From Kopecko, D. J., Brevet, J., & Cohen, S. N. (1976). *Journal of Molecular Biology* 108:333–60).

6 Recombination at the cointegrate resolution (res) site in the transposon, separates A and B.
7 B now has a copy of the transposon which is flanked by a duplication of a short sequence of B DNA.
(Redrawn from Shapiro, J. A. (1979). *Proceedings of the National Academy of Sciences* USA 76: 1933–7.)

containing antibiotics, consist of a single RTF unit linked to tandemly repeated r-determinants. It seems that the poly-r determinants are formed by transposition or recombination. The whole of the r-determinant can in fact transpose to other replicons. The r-determinant is incapable of maintaining itself in bacteria in the absence of the RTF. Whatever the mechanism of accumulation of poly-r determinants, the formation of exact duplications of the r-determinant suggests that the IS1 elements at the ends are involved. Molecules consisting entirely of repeated r-determinants are also found in *Proteus* and may be formed by recombination. When bacteria containing poly r-determinants are grown in drug-free growth medium for a few generations, almost all the R plasmids consist of a single RTF linked to a single r-determinant.

Mechanisms of antibiotic resistance

Plasmids make their bacterial hosts resistant to antibiotics by one of four mechanisms (see Foster, 1983):

1 Altering the target site of the antibiotic.
2 Modifying the antibiotic so that it is no longer active.
3 Preventing the antibiotic from entering the cell.
4 Specifying an enzyme which provides a substitute for a host-specified enzyme which is the target of the antibiotic.

Alteration of the target site: erythromycin and lincomycin-resistance Plasmids found in staphylococci and streptococci code for enzymes which bring about methylation of two adenine residues on 25*S* RNA molecules of bacterial ribosomes so that lincomycin and erythromycin cannot bind to them. This is the only known example of plasmid-mediated resistance brought about by a change in the target site of the antibiotic. The organism which is used to produce erythromycin, *Streptomyces erythreus,* has dimethylated 23*S* RNA, presumably to protect itself from the antibiotic it produces. It has been suggested that the *Streptomyces* enzyme which catalyses the methylation reaction might be the ancestor of the R plasmid enzyme.

Detoxification by modification of the antibiotic

a Chloramphenicol-resistance Chloramphenicol-resistance is brought about by plasmids which code for chloramphenicol acetyltransferase. This enzyme detoxifies chloramphenicol by catalysing the formation of 3-acetoxychloramphenicol.

Chloramphenicol + acetyl-S-CoA→3-acetoxychloramphenicol+
HS-CoA

Chloramphenicol acetytransferases are coded for by plasmids found in Gram-positive and Gram-negative bacteria. Several enzymes have been purified and parts of the amino acid sequence determined. Comparisons of these sequences show that the enzymes encoded by plasmids from various Gram-negative bacteria are related to each other, indicating that the enzymes have evolved from a common ancestor. The enzymes produced by Gram-positive bacteria are also related to each other, but the only similarity between the enzymes from Gram-negative and Gram-positive bacteria is a short sequence of amino acids at the chloramphenicol-binding site. It is not known whether this represents a residual low level of homology, indicative of common ancestry, or whether it arises because similar requirements in two unrelated enzymes have led to the selection of a similar sequence. Chloramphenicol acetyltransferase is an intracellular enzyme. It is expressed constitutively in *E.coli* and is inducible in *S. aureus*.

b Penicillin and cephalosporin-resistance Most strains of *S. aureus* are now resistant to penicillin G, the original penicillin discovered by Fleming which was initially used with such dramatic effects against this bacterium. But penicillin G is still useful for treating infections by many other Gram-positive bacteria. Modified forms of the original penicillin are now widely used because they have a greater spectrum of action or can be given orally. Ampicillin is a widely used penicillin derivative which can be used against many Gram-negative bacteria. *Pseudomonas aeruginosa* is comparatively resistant to ampicillin (even when it does not contain an R plasmid) but it is sensitive to another penicillin derivative, carbenicillin, and to cepharlosporins.

All these penicillins are inactivated by β-lactamases (penicillinases) produced by many Gram-positive and Gram-negative bacteria. The enzyme may be specified by chromosomal or plasmid genes. Some strains produce both types of enzyme. β-lactamases catalyse the hydrolysis of the β-lactam ring. They are classified into numerous groups according to their substrate specificities and electrophoretic mobilities. Most of the enzymes produced by Gram-negative bacteria belong to the TEM-1 or TEM-2 types (so-called because the first enzyme of this type to be analysed was specified by the plasmid R6K which was formerly called R-TEM). Both these types of β-lactamase are encoded by transposons, which no doubt contributes to their widespread distribution and their occurence on many unrelated plasmids. The plasmid-determined enzymes of Gram-negative bacteria are synthesized constitutively and accumulate in the periplasmic space. The β-lactamases produced by Gram-positive bacteria are usually inducible and extracellular.

These are numerous reports of bacteria which have become resistant to the new generation of cephalosporins (broad spectrum antibiotics related to the penicillins) because they produce β-lactamases. Production of β-lactamases is not the only mechanism of bacterial resistance to penicillins; for example, mutations can reduce the affinity of the cell's penicillin-binding proteins towards penicillin.

Preventing accumulation of the antibiotic

a Tetracycline-resistance *E.coli* accumulates tetracycline in two ways: by an energy-dependent (active) process and by a passive transport system. The energy-dependent uptake of tetracycline suggests that this antibiotic resembles a useful metabolite for which *E.coli* has evolved an active transport system. Plasmids conferring tetracycline-resistance promote the efflux of tetracyline. This is an energy-dependent reaction, but the biochemical mechanism for pumping tetracycline out of the cells is unknown. There is no evidence that tetracycline is modified by resistant bacteria.

Tn*10* is one of the transposons found in enterobacteria which codes for tetracycline-resistance. Tetracycline induces resistance when it is added to Tn*10*-containing cells at concentrations which do not greatly inhibit growth. Tn*10*

Table 5 Antibiotics and other antimicrobial drugs

Antimicrobial	Mechanism of action	Mechanism of plasmid-mediated resitance
Penicillins	Inhibit cell wall synthesis. Bactericidal.	β-lactamases hydrolyse β-lactam ring.
Penicillin G (Benzyl penicillin)		
Ampicillin		
Carbenicillin		
Methicillin		
Cephalosporins	Inhibit cell wall synthesis (like penicillins). Bactericidal.	β-lactamases.
Sulphonamides	Competitive inhibitors of dihydropteroate synthetase. Bacteriostatic.	Sulphonamide-resistant dihydropteroate synthetase specified by plasmid.
Trimethoprim	Inhibits dihydro-folate reductase. Bacteriostatic.	Trimethoprim-resistant dihydrofolate reductase

Tetracyclines	Protein synthesis (bind to ribosomes). Bacteriostatic.	Prevents intracellular accumulation
Chloramphenicol	Protein synthesis (binds to ribosomes). Bacteriostatic.	Detoxification by chloramphenicol transacetylase.
Aminoglycosides	Protein synthesis (cause misreading of mRNA). Bactericidal.	Modification of antibiotic which prevents uptake.
Streptomycin		
Kanamycin		
Gentamicin		
Amikacin		
Erythromycin	Protein synthesis (binds to ribosomes). Bacteriostatic.	Methylation of 23*S* rRNA of ribosomes.
Lincomycin	Protein synthesis (binds to ribosomes). Bacteriostatic.	Methylation of 25*S* rRNA of ribosomes.

encodes a protein which is found in the cell envelope and a second protein, found in the cytoplasm, which represses synthesis of the protein found in the envelope.

b Aminoglycoside-resistance The aminoglycoside group of antibiotics includes streptomycin, spectinomycin, kanamycin, gentamicin and amikacin. Plasmids confer resistance to aminoglycosides by coding for enzymes which modify them. The effect of this modification is to block the mechanism which normally transports the antibiotic into the cell. In addition, modified aminoglycosides are inactive against ribosomes.

Plasmid-determined enzymes which modify aminoglycosides are found in both Gram-positive and Gram-negative bacteria. They bring about *N*-acetylation, *0*-phosphorylation or *O*-nucleotidylation. The acetylating enzymes use acetylcoenzyme A as a co-factor. The nucleotidylating enzymes and phosphorylating enzymes use ATP or other nucleotides as substrates. The reactions catalysed by these enzymes may occur at one of several amino- or hydroxyl groups on aminoglycosides so that at least 12 types of enzyme can be distinguished according to their mode of action. The enzymes are found in the cell walls of Gram-negative bacteria and appear to be constitutive.

Enzyme substitution

a Sulphonamide-resistance Sulphonamides are bacteriostatic because they are competitive inhibitors of the enzyme dihydropteroate synthesis which is required for the synthesis of dihydrofolate. Plasmids found in Gram-negative bacteria code for a dihydropteroate synthetase which is about 1000 times less sensitive to sulphonamide than the parent enzyme. The enzyme is synthesized constitutively.

b Trimethoprim resistance This drug inhibits another enzyme involved in folate metabolism, dihydrofolate reductase. Sulphonamides and trimethprim are often used together as they act synergistically by inhibiting different enzymes in the same pathway for the synthesis of folate co-enzymes. Plasmids conferring trimethoprim-resistance in Gram-negative bacteria code for a dihydrofolate reductase which is about 20,000 times more resistant to trimethoprim than the chromosomal enzyme. The plasmid-encoded enzyme is a tetramer having a molecular weight of 35,000, in contrast to the *E.coli* chromosomally-specified enzyme which has a molecular weight of about 20,000. It has been suggested that the plasmid-encoded enzyme might have been derived from a T-phage enzyme which has a similar molecular weight.

Resistance to metals

Plasmids found in enterobacteria can confer resistance to the ions of arsenic, silver, copper, mercury and tellurium. Staphylococcal plasmids can confer resistance to arsenic, bismuth, cadmium, copper, lead, mercury and zinc compounds; *Pseudomonas* plasmids can confer resistance to chromium, mercury and tellurium ions. Resistance to ions such as Hg^{2+}, Ag^{+} or TeO_3^{2-} can be increased more than 100-fold by these plasmids.

Plasmid genes conferring resistance to mercuric ions are especially common. About 25% of conjugative R plasmids found in enterobacteria and about 75% of R plasmids from *Pseudomonas aeruginosa* confer resistance to Hg^{2+}. Plasmids from *P.aeruginosa* strains isolated from patients are more likely to have genes coding for Hg^+ resistance than genes for antibiotic resistance.

Resistance to mercuric ions is brought about by a plasmid-determined reductase which reduces Hg^{2+} to volatile Hg^0. This is insoluble in water and is rapidly released as a vapour when mercuric-resistant bacteria are grown in liquid media containing mercuric ions. The vapour can be collected in a condensor to yield liquid metallic mercury. Plasmids conferring mercuric-resistance also specify a mechanism for the uptake of mercuric ions. The genes for the reductase and for transport are part of an operon which is inducible by Hg^{2+}. Transposon Tn *501* (p.58) confers resistance to mercuric ions. Resistance to cadmium and arsenate are caused by plasmid-determined efflux mechanisms.

Plasmids conferring metal resistance are selected in environments containing metals in much the same way that R plasmids are selected by antibiotics. Plasmids conferring resistance to silver have been found in enterobacteria and in *Pseudomonas aeruginosa* strains isolated from patients with burns who have been treated with silver nitrate. Almost all the bacteria isolated from the sludge produced by industrial reprocessing of used photographic film were found to have conjugative plasmids conferring resistance to silver, mercury and tellurium ions. Plasmids conferring silver-resistance can also be found in bacteria isolated from silver mines.

Plasmids conferring copper resistance have been found in enterobacteria isolated from livestock which has been fed on feed-stuffs containing copper as a growth promoter.

Evolution of drug-resistant bacterial populations

In order to assess the implications of R plasmids for human and veterinary medicine, the following aspects of the ecology of R plasmids are considered here:

1 the increase in drug-resistance, especially that which has occurred in the last 30 years.
2 the nature of the selection pressures which have brought about the increase in drug-resistance.
3 recent examples of the problems posed by R plasmids, namely, their spread to new species, especially those which cause epidemics, and their occurrence in strains which cause infections in hospitals.

The threat to antimicrobial chemotherapy posed by the evolution of drug-resistance became apparent long before plasmids were discovered. Treatment of bacterial infections with drugs often selects resistant variants of the pathogen. Resistance may be the result of mutations in chromosomal genes or the presence of R plasmids. If the selection pressure provided by a drug is continued for many years, the resistant form of a bacterium can become the predominant type. Sulphonamides were successfully used in the 1930s to treat gonorrhoea, but by the late 1940s more than 80% of *Neisseria gonorrhoeae* strains were resistant to high levels of these drugs. *N.menigitidis,* enterobacteria and *Streptococcus pyogenes* also became largely sulphonamide-resistant. The role of plasmids in the evolution of sulphonamide-resistance in these species is not entirely clear. Plasmids specifying sulphonamide-resistance are now widespread in many enterobacteria but the resistance in many other bacteria appears to be due to mutations of chromosomal genes.

Penicillin-resistant strains of *Staphylococcus aureus* were rapidly selected after penicillin came into use in the 1940s. By 1946, 14% of strains in hospital where penicillin was extensively used were penicillin-resistant. This proportion rose to 38% in 1947 and to 59% in 1949. Almost all *S.aureus* strains isolated from hospitals now are penicillin-resistant and produce an inducible β-lactamase. In retrospect, it seems that R plasmids were responsible for much of the penicillin-resistance which became apparent in the late 1940s. As early as 1949, Barber noted that penicillin-resistance was often lost from staphyloccocci after they had been kept for a few months in the laboratory. Plasmids specifying β-lactamases have been found in many strains of *S.aureus* which have been isolated more recently.

In the last 25 years, the widespread use of a greater variety of antibiotics has been accompanied by an increase in multiple drug-resistance. The evolution of multiple resistance in *S.aureus* and in enterobacteria (especially *E.coli, Shigella* and Salmonella) has been particularly rapid. The numbers of drug-resistant strains of *Shigella sonnei* (a cause of a mild form of dysentery) isolated from patients in London increased rapidly during the period 1956–1967. Most of the resistance was plasmid-encoded. The antibiotic resistance of *Shigella dysenteriae, Shigella flexneri* and *Shigella boydii* isolated in England and Wales is shown in Fig 19. It has been pointed out that these strains of *Shigella* are not indigenous to Britain. Most of the infections are in people returning from the Indian

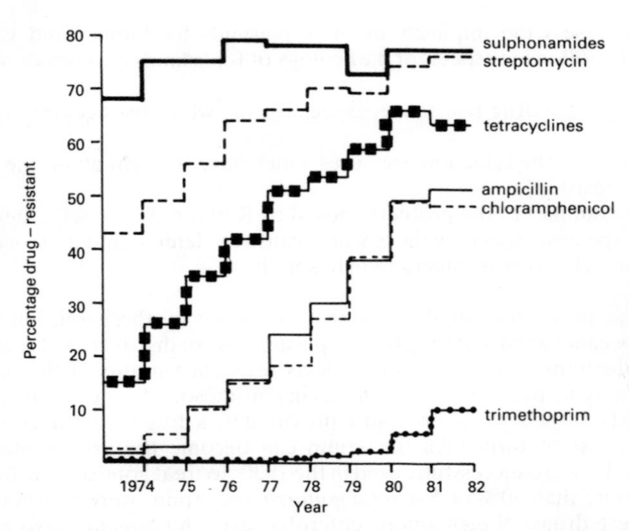

Fig. 19 *Shigella* drug resistance, 1974–1982 (From. Rowe, B. and Threlfall, E. J. (1984) *British Medical Bulletin* 40: 68–76).

sub-continent. In fact, multiple drug resistance in *Shigella sonnei*, which is indigenous to Britain, decreased from 38% in 1972 to 8% in 1977 (Rowe and Threlfall, 1984).

Resistant strains of *Salmonella* are common in many countries. The occurrence of large outbreaks caused by individual strains appears to be responsible for the considerable variation in the incidence of drug-resistant salmonellae from year to year (see Table 6).

Uses of antibiotics which have selected R^+ bacteria

Antibacterial drugs are used to treat infections in man and animals. They are also used for prophylaxis (to prevent infections) and as growth promoters when added to food eaten by animals. The large amounts of drugs given to livestock (either for prophylaxis or for growth promotion) has raised the question of whether these practices select R^+ bacteria which could seriously complicate the therapy of animal and human infections.

Selection of R^+ bacteria in people When patients are given oral doses of tetracycline the predominant strains of *E.coli* recovered from their faeces are usually tetracycline-resistant within a week. If the treatment is continued for only a short period, the number of antibiotic-resistant *E.coli* in the alimentary tract usually falls back to a low level after two or three weeks: sometimes an antibiotic-resistant strain selected by the period of treatment may continue to be the predominant type for several months after treatment with the antibiotic has stopped. Bearing in mind that at least one third of people admitted to hospitals receive some form of antimicrobial drug during the course of their stay, it is clear that there is a considerable selection pressure for drug-resistant bacteria in hospitals. In addition, 10 times as many prescriptions for antibiotics are dispensed

Table 6 Variation in antibiotic resistance in *Salmonella* strains isolated from domestic animals in the Federal Republic of Germany

	Year of isolation					
Type	1971	1972	1973	1974	1975	1976
	Percentage resistant					
S. typhimurium	32	37	64	52	36	36
S. panama	65	58	52	22	6	2
S. dublin	3	1	7	16	70	70

(From *Surveillance for the prevention and control of health hazards due to antibiotic-resistant entrobacteria.* (1978). Geneva: World Health Organization.)

to out-patients. In 1975 there were 19 800 000 prescriptions for penicillins and 11 100 000 for tetracyclines dispensed in the UK to people who were not in hospital. A greater proportion of *E.coli* strains isolated from hospital sewage are multiply drug-resistant in comparison with strains from domestic sewage. However, analyses of *E.coli* strains from all sources entering the Bristol sewage works indicated that more than 90% of all the enterobacterial R plasmids in pooled sewage came from people who were not in hospital.

It has been argued that many prescriptions for antibiotics are unnecessary. They are sometimes used inappropriately to treat viral infections or mild gastroenteritis. Treatment of *Salmonella* gastroenteritis in man or animals with antibiotics does not aid recovery but merely prolongs the period of excretion of salmonellae.

The rapid increase in the proportion of resistant bacteria in people who have been treated with antibiotics is due mainly to the selection of R^+ bacteria which were either already part of their bacterial flora or were ingested in food soon after treatment began. Transfer of R plasmids between bacteria in the alimentary tract can be demonstrated, particularly when antibiotics are given to select R^+ recipients, but transfer of conjugative R plasmids in the gut is not responsible for the rapid increase in the numbers of R^+ bacteria following antibiotic treatment.

Although the proportion of R^+ *E.coli* in people not receiving antibiotics is low (usually less than 0.1%), R^+ strains can nevertheless be demonstrated in at least 50% of people. These are the bacteria, which almost all of us have living inside us, that are selected when we take antibiotics.

Selection of R^+ bacteria in animals Antibacterial drugs are incorporated in animal feedstuffs to be used for the treatment of infections, for prophylaxis or for growth promotion. In many countries, low doses of tetracycline, sulphonamides and penicillin have been used in livestock feedstuffs since the 1950s to enhance the growth of animals, particularly poultry and pigs. It was discovered in 1948 that the growth rate of chickens increased when they were fed low doses of antibiotics such as tetracyclines. The physiological basis of growth promotion by antibiotics is still unclear.

Far more drug-resistant *E.coli* are present in the alimentary tracts of pigs and poultry which have been fed low doses of antibiotics. Most of the resistance is plasmid-specified. Drug-resistant strains of *S.aureus* and *Clostridium perfringens* are also much more commonly found in animals which have been fed antibiotics. The high level of resistance to penicillin and tetracycline amongst strains of *S.aureus* causing staphylococcosis in fowls fed with these antibiotics has meant that these drugs have become virtually useless for treating this disease.

The practice of including tetracycline in the food eaten by broiler fowls in the UK began in about 1956 and then increased rapidly. During this period, the proportion of tetracycline-resistant strains of *E.coli* (isolated from fowls) belonging to commonly pathogenic serotypes greatly increased. The percentages of resistant *E.coli* were 3.5% in 1957, 20.5% in 1958, 40.9% in 1959, and 63.2% in 1960.

The obvious problem associated with the extensive use of antibiotics as feed additives is that it may select large numbers of drug-resistant bacteria so that the drugs become less effective when used to treat infections. The reduced effectiveness of antibiotics becomes an immediate problem for the treatment of bacterial infections in livestock, but several of the drugs added to livestock feedstuffs are also used to treat infections in people. Enterobacteria (and their R plasmids) are transmitted from animals to man, so the selection of R^+ bacteria in animals could also complicate the treatment of human diseases.

A UK Government committee, chaired by Professor Swann, was set up to assess this problem. One of their recommendations (Swann Report, 1969) was that drugs which are useful for treating infections in man or animals or which select resistant strains which could impair the efficiency of theraputic drugs should not be used as feedstuff additives for growth promotion. This recommendation was implemented in 1971. Similar measures have been adopted in other countries, including the Federal Republic of Germany, Ireland, Czechoslovakia and the Scandinavian countries.

In the U.S.A., antibiotics such as penicillin and tetracycline have not been banned from feedstuffs, although a proposal to prohibit them is being actively debated.

Following the implementation of the Swann proposals, analyses of *E.coli* isolated from pigs showed that the numbers of resistant bacteria had decreased only slightly. Samples taken from 100 pigs in different pens at Chelmsford market in 1975 showed that all were excreting tetracycline-resistant bacteria. The incidence of transmissible tetracycline-resistance had decreased more than the incidence of non-transmissible (possibly chromosome-determined) resistance. The persistence of tetracycline-resistance in these animals indicates that prolonged selection by a drug may lead to the evolution of strains which can compete effectively with drug-sensitive strains.

Tests made in 1980 showed that there was still a high proportion of antibiotic resistant *E.coli* strains in pigs and chickens. Transmissible tetracycline resistance was found in 68% of chicken strains and in 20% of pig strains. However, a smaller proportion (17.9%) of strains causing generalised infections in chickens were tetracycline resistant in 1982 than in 1972 (31.2%). The incidence of resistance to furazolidine and to sulphonamide had also decreased.

One of the difficulties in interpreting these results is that it is not clear whether the exposure of animals to tetracycline had decreased significantly since 1971. Feedstuffs containing tetracycline are sold on prescription for the treatment of infections and for prophylaxis. In addition, there have been several cases in the U.K. involving the illegal sale or importation of antibiotics (chlortetracycline, penicillin and sulphadimide) in which the defendants were found guilty.

Animal-to-man transmission of R^+ bacteria

One of the reasons for implementing the recommendations of the Swann Report was that the transfer of R^+ bacteria from animals to man might complicate the treatment of human infections. Transfer of R^+ pathogenic bacteria could create immediate difficulties, but the transfer of commensal R^+ bacteria such as *E. coli* could also complicate matters in the long run because R plasmids may be transferred to pathogenic strains in the human alimentary tract.

Animal carcasses, especially poultry, are often heavily contaminated with R^+

E. coli which can colonize the human alimentary tract. The handling of raw meat is a major route for transmission of strains from animals to man. Bacteria on raw meat may contaminate surfaces and utensils used for preparing food and may be spread to other food. Studies of the intestinal flora of hospital patients also indicate that contaminated food is one of the major routes for the transmission of *E. coli* strains which subsequently colonize the alimentary tract. In addition, animals form the major reservoir for strains of *Salmonella* (apart from *Styphi*) which are responsible for many cases of food poisoning.

The spread of an R^+ strain of *S.typhimurium* from animals to man during the period 1963–1966 has been well-documented. (Salmonellae cause numerous infections in livestock, especially in calves and poultry, and are an important cause of food poisoning and other diseases in man.) During 1964–1966 the intensive system of calf-rearing came into use in many farms in the UK. Because of the debilitating effects of poor animal husbandry on some of these farms, the calves became pre-disposed to *Salmonella* infections and to other diseases. In severe outbreaks of salmonellosis, mortality rates amongst calves can be as high as 50%. The nature of the intensive system of calf-rearing is such that the animals are moved from one part of the country to another for rearing in different farms at various stages of their development. This gave ample opportunity for the outbreak of salmonellosis to spread. Antibiotics were used extensively in attempts to control the outbreak and for prophylaxis to prevent other animals from becoming infected, but, in the absence of improvements in the animals' living conditions, therapy was often unsuccessful and losses continued to be high.

The outbreak of salmonellosis in cattle was caused by a phage type 29 strain of

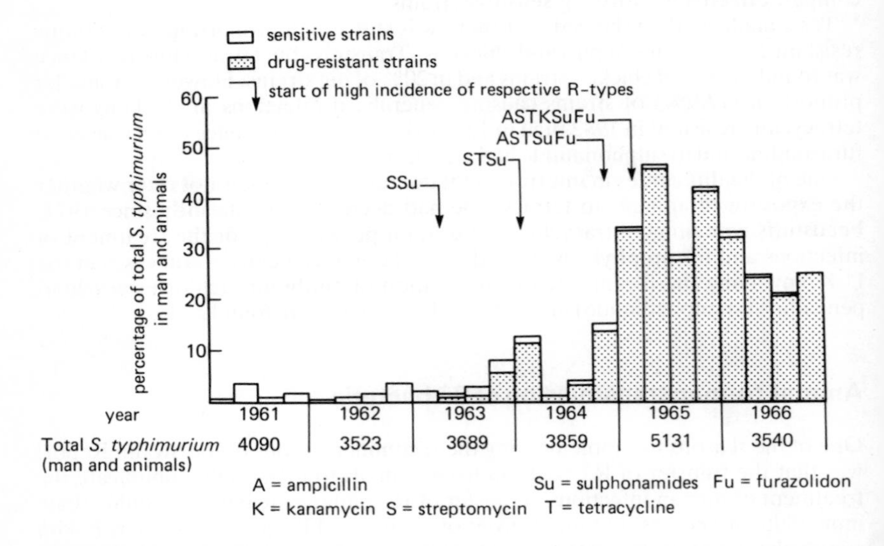

Fig. 20 Type 29 of *Salmonella typhimurium* as a percentage of total human and animal *S. typhimurium* infections. 1961–1966. (From Anderson, E. S. (1968. *Annual Review of Microbiology* 22: 131–80))

S. typhimurium. During the course of this outbreak this strain acquired resistance to ampicillin, streptomycin, kanamycin, tetracycline, sulphonamides and furazolidine (Fig.20). Resistance to all but furazolidine was coded for by R plasmids. In 1965, 576 strains of this type isolated from human infections were analysed and 555 of these were drug-resistant. There were six deaths caused by the strain.

The outbreak of salmonellosis caused by the type 29 strain was largely over by 1970. The predominant strain of *S. typhimurium* causing infections can change rapidly. By 1979, the most common *S. typhimurium* strain causing infections in cattle and in man in the U.K. was type 204. Again, the extensive use of antibiotics in attempts to control the spread of the disease appears to have been responsible for the selection of a multiresistant R^+ strain, first detected on a farm in Leicestershire. Infected calves have been sent to many parts of the country so that infections caused by the resistant type 204 are now widespread. More than 600 human infections and several deaths caused by multiply-resistant strains of type 204 were reported up to 1983.

Although the widespread use of antibiotics in agriculture has greatly reduced the effectiveness of several drugs used for treating animal infections, it is more difficult to assess the extent to which agricultural uses of antibiotics have contributed to the numbers of R^+ strains which cause human infections. R^+ bacteria are certainly transferred to man from animals. Some of them can cause serious diseases in man.

It should be borne in mind, that in the U.S.A. 50% of all antibiotics produced are used to promote animal growth and 50% in human and veterinary medicine. A total of nineteen antibiotics, including tetracycline and penicllin, are approved for growth promotion.

Problems caused by R plasmids in human pathogens

The rapid increase in the numbers of R^+ *Salmonella* strains during the 1960s indicated that plasmids might also be transferred to the highly virulent bacterium *Salmonella typhi*. Resistance to chloramphenicol would be especially serious because this is the most effective drug for treating cases of typhoid fever. In view of the large numbers of R plasmids circulating among the enterobacteria, the first typhoid epidemic to be caused by a chloramphenicol-resistant strain, in Mexico in 1972–73, was not a wholly unexpected event.

The strain contained a conjugative plasmid coding for resistance to chloramphenicol, streptomycin, sulphonamide and tetracycline. Presumably, the plasmid was formed in other enterobacteria exposed to these drugs and was then transferred to *S. typhi*.

Conditions exisiting in Mexico were considered to have increased the probability that a chloramphenicol-resistant strain of *S. typhi* would arise. There was a relatively high incidence of typhoid fever in the country, so *S. typhi* would often be present in alimentary tracts together with R^+ bacteria which could act as donors. In addition, antibiotics were available without the need for a doctor's prescription, thus increasing the probability of indiscriminate use.

There were at least 10 000 cases in the Mexico epidemic and many deaths occurred, especially during the first few weeks before ampicillin was used.

Outbreaks of typhoid fever caused by chloramphenicol-resistant strains have also occurred in India, Vietnam and Thailand.

Instances have been reported where an initially sensitive *S. typhi* has received an R plasmid during the course of an infection, presumably from another bacterium in the patient.

A multiply-resistant strain of *Shigella dysentariae* type 1 caused a large outbreak of dysentery in Central America from 1968 to 1972. Hundreds of thousands of people were affected and there were tens of thousands of deaths. The plasmid-determined resistance of the strain delayed diagnosis of the disease in some cases, since it did not respond to antibiotic therapy and was thought to be caused by *Entamoeba*. Another large outbreak caused by resistant *S. dysenterieae* occurred in Bangladesh in 1973.

Strains of *Vibrio cholerae* resistant to several antibiotics, including tetracycline, have caused several protracted outbreaks of cholera. Tetracycline is the drug of choice when antibiotics are indicated for cholera patients.

Recent examples of the spread of R plasmids to previously antibiotic-sensitive genera Penicillin is the antibiotic principally used for treating gonorrhoea, and ampicillin is the drug of choice for meningitis and pharyngitis caused by *Haemophilus influenzae*. β-lactamase-producing strains of *H.influenzae* were first isolated in 1974 and strains of *Neisseria gonorrhoeae* producing the enzyme were detected for the first time in 1976.

The β-lactamase produced by resistant strains of *H.influenzae* is of the TEM type and is coded for by plasmids, some of which are conjugative. A conjugative plasmid from *H. influenzae* was found to carry a transposon similar to Tn*1* and Tn*3* which code for β-lactamase.

Conjugative plasmids specifying resistance to kanamycin, chloramphenicol or tetracycline have also been found in *H. influenzae*. In one case, tetracycline-resistance was transposable and in another instance resistance to both tetracycline and chloramphenicol was transposable as a single unit. If multiresistant strains of *H. influenzae* become predominant, the treatment of the life-threatening disease of meningitis caused by this organism will become very difficult. The incidence of ampicillin resistance amongst *H. influenzae* strains isolated in the U.S.A., increased from 18.7% in 1978 to 23.9% in 1982. Two antibiotics (ampicillin and chloramphenicol) must be given initially to patients until the antibiotic-resistance spectrum of the strain has been determined.

The first two β-lactamase-producing strains of *N. gonorrhoeae* were isolated in 1976 from people who became infected in the Far East or West Africa. The two isolates had non-conjugative plasmids which specified a TEM-type β-lactamase. The plasmids shared much DNA sequence homology with each other and with a small β-lactamase-specifying plasmid found in *H. influenzae*.

About 7% of gonococci also have a conjugative plasmid which can mobilize the transfer of the small β-lactamase-specifying plasmid to other strains of *N. gonorrhoeae* or to *E. coli*. The small number of penicillinase-producing gonococci isolated in the UK each year are from people who became infected abroad; the resistant strain has not yet become endemic in most parts of the world.

The isolation of an ampicillin resistant strain of *Neisseria meningitidis* was first reported in 1983. The strain harboured a small plasmid specifying β-lactamase and a larger conjugative plasmid which appeared to be identical to plasmids found in *N. gonorrhoea. N. meningitidis* is responsible for most cases of meningitis and can cause epidemics of the disease. In the absence of effective chemotherapy (ampicillin is the drug of choice) the disease is usually fatal.

R plasmids in strains causing infections acquired in hospitals Many patients who are admitted to hospitals are suffering from infectious diseases so that the causal organisms inevitably contaminate the hospital environment. Other patients are predisposed to bacterial infections because they are already suffering from bacterial or viral infections; or their resistance to infection may be lowered because they are very old, very young or are being treated with immunosuppressive drugs. The introduction of catheters to drain the bladder causes many urinary tract infections. Enterobacteria, predominantly *E. coli,* are responsible for most urinary tract infections, the most common type of infection acquired in hospitals. About a half of hospital-acquired infections are caused by Gram-negative bacteria, primarily enterobacteria and *Pseudomonas; S. aureus* is the most important pathogen among the Gram-positive bacteria.

Chemotherapy of hospital-acquired infections is often complicated by the multiple drug-resistance of the pathogens involved. Valuable time may be lost if the first drug to be administered subsequently proves to be ineffective. It is often found that effective drugs which have few side effects cannot be used because of plasmid-determined resistance.

Some of the most serious and widespread outbreaks in recent years particularly in developing countries, have been caused by drug-resistant strains of *Salmonella.* These have occurred in many parts of the world and fatality rates in paediatric units and nurseries can be as high as 25%. Several of these strains have been shown to harbour virulence plasmids (see p.82) The *Salmonella* strains are almost always resistant to several antibiotics. This can greatly complicate the treatment of septicaemia and meningitis which frequently occur in these outbreaks.

Another example of the problems which can arise in the treatment of hospital-acquired infections is provided by the experience of a Burns Treatment Unit at the Birmingham (UK) Accident Hospital. Their work illustrates that the problem of antibiotic-resistance can be contained, but only by the most careful monitoring of strains and prudent use of antibiotics.

Extensive burns are initially sterile but can quickly become colonized by *Pseudomonas aeruginosa* and a variety of enterobacteria which can cause serious infections. *P. aeruginosa* can be a particularly troublesome pathogen because it is inherently resistant to all but a few drugs, so the acquisition of an R plasmid can narrow the choice to perhaps one or two drugs.

Carbenicillin was particularly effective against *P. aeruginosa,* but the use of this antibiotic in the Burns Unit led to the emergence of a resistant strain which rapidly replaced the sensitive form previously present in the hospital. In an attempt to eliminate the resistant pseudomonad, use of carbenicillin was stopped for a while so that there was no selection for the strain. Although this procedure led to the removal of the strain from the ward, when treatment of patients with carbenicillin was resumed the resistant pseudomonad quickly reappeared.

Analyses of strains present in the wards showed that the conjugative plasmid conferring carbenicillin-resistance was also present in other species of bacteria, including a strain of *Klebsiella.* When treatment of *Pseudomonas* infections with carbenicillin was resumed, R^+ recipients of the pseudomonads were selected.

The plasmid responsible was RP4 (RP1 and RK2 were isolated from the same hospital and are probably the same as RP4) which has a wide host range and specifies resistance to tetracycline, kanamycin and chloramphenicol as well as a β-lactamase (coded for by a transposon, Tn*1*). The use of antibiotics other than carbenicillin did not select R^+ forms of *P. aeruginosa* as this species is inherently resistant to tetracycline and kanamycin, However, *Klebsiella* is normally sensitive to tetracycline and kanamycin, so the R^+ forms of this bacterium persisted in the hospital to provide a reservoir of R plasmids which could be transferred to *Pseudomonas*. Strains harboring the R plasmid could be eliminated only by withholding the use of all antibiotics which had corresponding resistance determinants on the plasmid. The adoption of this policy, together with the discharge of certain patients from the ward, eventually succeeded in removing the R plasmid.

Gentamicin is one of the few antibiotics which can be used to treat infections caused by carbenicillin-resistant *P. aeruginosa,* but an outbreak of infections caused by a gentamicin-resistant strain recently occurred in a Burns unit. The resistance was plasmid-specified. In this case, the incidence of the resistant strain in the ward could not be reduced simply by withholding use of gentamicin, but only by segregating all patients with *P. aeruginosa* infections in one ward and admitting all new patients to a separate ward to prevent cross-infection.

Evolution of R plasmids

R plasmids can be considered to be composed of two major elements which evolve as more or less separate entities. These are the essential replication region (the origin of replication and any genes and sequences required for replication and segregation) and the drug-resistance genes, many of which are transposons. Some transposons can be separated into two components—the genes specifying resistance and the insertion sequences which are responsible for transposability. These various components of plasmids may evolve separately in the sense that each is a *selfish gene* in its own right. More or less permanent associations come about which are to the mutual selective advantage of the individual components.

Speculation about the origin of the replication region of bacterial plasmids has centred on the possibility that fragments of bacteriophages or bacterial chromosomes could give rise to plasmids. A small fragment comprising the *oriC* region of the *E. coli* chromosome can in fact replicate as a circular plasmid in this bacterium. It is maintained at about the same copy number as the chromosome.

We can only guess at the origins of insertion sequences. Are they perhaps derived from genes coding for restriction endonucleases or from enzymes involved in general recombination which have evolved to recognize specific sequences? Or are they highly defective bacteriophages of the type represented by Mu, which seems to transpose in much the same way as IS-elements? If this is the case then all that is left are the ends of the phage chromosomes and perhaps a genes specifying a *transposase* which is necessary for recognition of the ends and for transposition.

One view of insertion sequences is that they are essentially parasitic genes; they simply promote their own replication and dispersal. Such genes will propagate themselves providing their activities do not greatly disrupt the functioning of other genes. This may be one of the factors which sets a limit to the number of IS-elements on a chromosome or a plasmid and lead to the evolution of a particular frequency of transposition. The optimum is that which leads to greatest propagation in the absence of inactivation of essential genes. In addition, recombination between multiple copies of IS-elements would lead to instability.

An alternative view is that the genetic rearrangements caused by IS-elements are a major factor in bacterial evolution: IS-elements have evolved and continue to survive because they generate variants enabling their bacterial hosts to become rapidly adapted to different environments. They cause deletions, additions, duplications, transpositions, inversions and fusions of genes, providing the raw material for evolution by natural selection. The two views are not incompatible; initially parasitic genes may have since acquired other properties.

Possible ancestors of R plasmid resistance genes are chromosomal genes which have mutated to antibiotic-resistance or genes from antibiotic-producing bacteria which have resistance genes to ensure they are not killed by the antibiotic they produce. The similarity between the mechanism of plasmid-determined erythromycin-resistance and that determined by the organism which produces erythromycin has been mentioned (p.64). The enzyme specified by a penicillin-resistance transposon was found to be similar to a β-lactamase coded for by a chromosomal gene in many strains of *Klebsiella.* It is, of course, difficult in such cases to determine which is ancestor and which is descendant.

Some transposons consist of a pair of IS-elements associated with a resistance gene. Given the prior existence of insertion sequences and of genes specifying antibiotic-resistance, the evolution of transposons seems almost inevitable. Insertion sequences which happen to become associated with drug-resistance genes would have a considerable selective advantage in many environments; an antibiotic-resistance gene would seen an ideal partner for any ambitious insertion sequence. Resistance is determined by a relatively small amount of DNA, it specifies a crucial gene-product in many environments and is expressed in many different bacteria. Instead of being a sequence which simply selects itself by being good at replication and transposition, an IS-element associated with a resistance determinant has a highly selectable phenotype. In the same way, plasmid replicons and transposons become associated to their mutual selective advantage. In the presence of antibiotics, a plasmid which carries a transposon is more likely to survive and propagate than one which does not.

The view is often expressed that R plasmids have in some way been selected by clones of cells so that most of the clone can dispense with the plasmid when it is not needed (that is, when there are no antibiotics in the environment). When the selection pressure is re-applied, the few cells in the clone which still have a copy of the plasmid will be selected. Plasmids may well be lost unless a selection pressure is applied, but this general view of R plasmids as dispensable cell elements seems to me to be unreasonable and to place undue emphasis on cells rather than genes as the fundamental elements of evolution. I doubt whether cells have evolved a mechanism for regulating R plasmid loss. R plasmids can be transferred to many different species and the rates of loss vary considerably. The view of R plasmid evolution proposed here places the emphasis on the selection of genes.

Whatever view is taken about the course of R plasmid evolution, it should not be assumed that all R plasmids have evolved recently, that is, since the use of antibiotics on a large scale. R plasmids specifying resistance to tetracycline and streptomycin have been found in strains of *E. coli* which were freeze-dried in 1946, before the clinical use of these antibiotics. They have been found in strains isolated from inhabitants of North Borneo and the Solomon Islands who had never been treated with antibiotics and presumably had had little, if any, contact with people or animals who had been so treated. R plasmids have also been found in soil organisms from isolated parts of the world. It has been suggested that they might evolve in soil as a response to the antibiotics produced by actinomycetes and other soil organisms.

R plasmids specifying resistance to new drugs can be detected shortly after their introduction to medicine. R plasmids coding for resistance to trimethoprim and to gentamicin were first detected in 1972, about three years after the introduction of these drugs. Trimethoprim is a synthetic drug; a natural counterpart, which might have selected an R plasmid before the drug was used in medicine, has not yet been found.

Summary

R plasmids were discovered in strains of *Shigella* and *Escherichia coli* isolated in Japan. They have since been found in a wide range of bacteria, both Gram-positive and Gram-negative. They can occur in almost all the major pathogens of man and animals and can confer resistance to almost all the antibacterial drugs used in human and veterinary medicine. R plasmids often code for resistance to three or more drugs.

Most R plasmids found in Gram-positive bacteria are non-conjugative, but conjugative R plasmids are common in many Gram-negative bacteria.

F-like R plasmids are common in the enterobacteria. The similarity between F and F-like plasmids can be shown by the genetic complementation which occurs between most of their *tra* genes and by comparisons of their DNA sequences using the heteroduplex technique.

Most of the drug-resistance genes of F-like R plasmids are clustered within a region of the plasmid called the *r-determinant* which has a copy of the IS1 insertion sequence at each end. The remaining part of the plasmid is called the *resistance transfer factor* or RTF and comprises all the genes required for replication and for conjugation

Many genes conferring drug-resistance are parts of special genetic elements known as *transposons*. These resemble insertion sequences in that they can transpose copies of themselves to other DNA molecules independently of the mechanisms which promote recombination between homologous DNA molecules in bacteria. Transposons range in size from about 2000 to 19,500 base pairs.

At least one third of patients admitted to hospitals receive antibiotics, and this is reflected in the high proportion of drug-resistant bacteria isolated from hospitals. However, about 90% of the R^+ *E. coli* in pooled sewage from a city came from domestic sources presumably because substantially more antibiotics are prescribed for people who are not in hospital than for hospital patients.

The major reason for the rapid increase in R^+ bacteria which usually accompanies antibiotic therapy appears to be the selection of pre-existent R^+ strains, rather than the rapid spread of R plasmids to previously sensitive stains which the patient is harbouring.

Many of the antibiotics which are used to treat human infections are also used to combat infections in animals, either for prophylaxis or to promote the growth of livestock. For reasons which are not altogether clear, the inclusion of low concentations of antibiotics in the food eaten by animals increases their growth rate. Some antibiotics have become useless for treating infections in animals because resistant strains have been selected by the use of antibiotic-containing feedstuffs.

The problem of bacterial drug-resistance became apparent soon after the introduction of sulphonamides in the 1930s. Although chromosomal drug-resistance can give rise to clinical problems, most of the current difficulties caused by multiply resistant bacteria arise because the strains contain R plasmids. R^+ bacteria are passed from animals to man; large numbers of enterobacteria often contaminate poultry and other meat. One of the major recommendations of a UK Government Committee was that antibiotics which are used to treat human infections should not be used as growth promoters in animal feedstuffs. Since the introduction of this legislation in 1971, there is evidence that the number of antibiotic resistant bacteria found in livestock is decreasing.

R plasmids cause the greatest problems when they occur in bacteria which cause major epidemics (for example, *Salmonella typhi* and *Shigella dysenteriae*) or in strains which cause infections in hospitals (enterobacteria, *Pseudomonas aeruginosa* and *Staphylococcus aureus*).

References

DATTA, N. (ed.) (1984). Antibiotic resistance in bacteria. *British Medical Bulletin* volume 40. No 1.

FALKOW, S. (1975). *Infectious Multiple Drug-Resistance*. Pion, London.

FOSTER, T. J. (1983). Plasmid-determined resistance to antimicrobial drugs and toxic metalions in bacteria. *Microbiological Reviews* 47:361–409.

GRINDLEY, N. D. F. and REED, R. R. (1985). Transpositional recombination in prokaryotes. *Annual Review of Biochemistry* 54:863–896.

LINTON, A. H. and HINTON, M. H. (1983). pp 533–549. In. *Antimicrobials and Agriculture*. Edited by M. Woodbine. Butterworths, London.

ROWE, B. and THRELFALL, E. J. (1984). Drug resistance in Gram-negative aerobic bacilli *British Medical Bulletin* 40: 68–76.

SHARP, P. A., COHEN, S. N. and DAVIDSON, N. (1973). Electron microscope heteroduplex studies of sequence relations among plasmids of *Escherichia coli*. II Structure of drug-resistance (R) factors and F factors. *Journal of Molecular Biology* 75: 235–55.

SMITH, H. WILLIAMS (1978). Antibiotic resistance in bacteria and associated problems in farm animals before and after the 1969 Swann Report. pp. 345–57. In: *Antibiotics and Antibiosis in Agriculture*. Edited by M. Woodbine. Butterworths, London.

5 Other plasmids

Virulence plasmids

Pathogenic bacteria are able to withstand host-defence mechanisms and often produce toxins which damage the host. Both of these properties are sometimes plasmid-specified. Highly potent toxins can also be encoded by bacteriophage chromosomes. The classical example of this is diphtheria toxin which was shown by Freeman in 1950 to be produced only when *Corynebacterium diphtheriae* is lysogenized by a particular bacteriophage. The structural gene for the protein toxin is on the bacteriophage chromosome.

Virulence plasmids of *E. coli* and *Shigella*

a Enterotoxigenic strains *E. coli* strains can cause a variety of different types of disease. The most common in man are urinary tract infections, but *E. coli* is also a major cause of diarrhoea. At least three types of strain causing diarrhoea can be distinguished. Some strains cause a disease resembling dysentery which is caused by *Shigella* (see below); epithelial cells of the colon are invaded and destroyed so that the faeces become mixed with blood. Other strains cause diarrhoea with little or no accompanying invasion of mucosa. The *enterotoxigenic strains* fall into this category. Many of these strains harbor plasmids (or sometimes temperate bacteriophages) which code for one or more types of enterotoxin. Enterotoxigenic strains are a common cause of acute diarrhoea in young animals, and in adults can cause *traveller's diarrhoea* which affects many visitors to tropical countries. A third class cause diarrhoea without extensive tissue invasion, but do not produce detectable enterotoxins. Perhaps they produce types of toxin which we are not yet able to assay.

Two kinds of toxin, heat-labile toxin (LT) and heat-stable toxin (ST), may be produced by enterotoxigenic strains. Many plasmids code for both types of toxin The heat-labile toxin has been studied most extensively. It comprises two protein sub-units; sub-unit A has a molecular weight of 25 500 and B has a molecular weight of 11 000. Each A sub-unit is associated with four or five B sub-units. Both sub-units are related to the corresponding parts of the cholera toxin produced by *Vibrio cholerae*. Sera raised against cholera toxoid will cross-react with LT and neutralize it. Cholera toxin and LT have clearly evolved from a common ancestor but comparisons of the DNA sequence coding for the two proteins show that there are numerous differences between them. Cholera toxin is encoded by chromosomal genes of *V. cholerae*.

Both LT and cholera toxin act in a similar manner on sensitive mammalian cells. The B sub-unit of LT binds to glycolipid receptors (ganglioside GM_1) in the cytoplasmic membrane. The A sub-unit of the toxin then penetrates the membrane and activates adenylate cyclase. It is known that this activation is by a mechanism which involves ADP-ribosylation of a protein found on the inne surface of the membrane. NAD is a co-factor in the reaction. *E. coli* LT probably acts in a similar manner. Stimulation of adenylate cyclase increases the concentration of 3′5′-cyclic adenosine monophosphate (cyclic AMP) which in turn causes water and electrolytes to be secreted from the cells into the lumen of the intestine.

The heat-stable toxin of *E. coli* is unrelated to LT. ST is a low molecular weight (about 5000) peptide and is only very weakly, if at all, antigenic. It also differs from LT in its effects on sensitive cells; ST stimulates guanylate cyclase, rather than adenylate cyclase. LT can be assayed by its effects on the shape of tissue culture cells, for example. Y1 adrenal cells *round-up* when they are treated with LT. ST cannot be assayed in this way but can be assayed by measuring the accumulation of fluid in the intestines of infant mice following the instragastric injecton.

A plasmid gene coding for ST has recently been shown to be part of a transposon (Table 4). The transposon comprises two IS*1* elements in opposite orientations; the gene coding for ST lies in the sequence of 500 base pairs which separates the two elements. The gene coding for LT has also been found on a temperate bacteriophage chromosome. Lysogenization of strains by the phage converts them to LT-producers. Treatment of the lysogenic strains with mitomycin C induces the prophages and stimulates LT production. Perhaps the LT gene is also part of a transposon and so may become a part of other replicons besides plasmids.

Enterotoxigenic strains do not achieve their full pathogenic potential unless they are also able to colonize the small intestine effectively. The colonization factors which are commonly found on enterotoxigenic strains are protein pili (also called fimbriae) which stick the bacteria onto the wall of the small intestine. In healthy animals, *E. coli* is confined almost exclusively to the large intestine. The adhesive pili which adhere to the wall of the small intestine are often specified by plasmids. These usually differ from the plasmids which code for enterotoxins.

The colonization factors show some specificity for particular animals. For example, almost all the enterotoxigenic strains isolated from pigs have adhesive pili which are either K88 antigen or 987P antigen. These antigens enable bacteria to colonize the small intestines of pigs effectively but have no effect on their ability to bind to the small intestine of calves, for example. Enterotoxigenic strains isolated from calves or from lambs usually produce a plasmid-specified K99 antigen, and human strains often have the colonization factor antigens CFAI or CFAII, which are also plasmid-determined. In addition to plasmid-determined features, many chromosomal gene-products are also important in determining whether strains can colonize various hosts or cause disease. Certain lipopolysaccharide O antigens and capsular polysaccharide K antigens occur more commonly amongst strains causing diarrhoea than on other strains. Strains causing diseases in pigs usually belong to particular serotypes, whereas those causing disease in calves belong to other groups. The transfer of virulence plasmids into a strain which is poorly adapted for life in the intestine, such as *E. coli* K-12, does not convert it into the type of pathogenic *E. coli* which can be isolated from diseased livestock.

The importance of the combined effects of enterotoxin and K88 antigen in causing diarrhoea of piglets is illustrated by the experiments of Smith (1976) who transferred various plasmids into a strain to test their effect on its virulence. The results show that the presence of the enterotoxin plasmid causes severe diarrhoea in piglets only when it is accompanied by a plasmid coding for K88 antigen, which enables the bacteria to adhere to the wall of the small intestine.

In view of the cooperative effects of colonization factors and enterotoxins in causing diarrhoea, it is perhaps surprising that both characteristics are rarely specified by the same plasmid. Genes coding for raffinose utilization are almost always present on plasmids coding for K88 antigen. Direct repeats of the insertion sequence IS*1* flank the genes for raffinose utilization and K88 antigen. Both properties can be transposed together. It has been suggested that Raf plasmids may enable pathogenic strains to produce a cell wall which is more resistant to host-defence mechanisms. Plasmids coding for haemolysins are much more commonly found in strains causing diarrhoea in pigs than in strains from other sources, indicating that the haemolysins may also contribute to virulence. Experiments to test this hypothesis do not indicate that haemolysin production increases the virulence of porcine strains causing diarrhoea. However, there is evidence which indicates that haemolysin production increases the virulence of strains when tested in animals for their ability to cause pyelonephritis or peritonitis.

The crucial role played by colonization factors has been exploited in the development of vaccines which are designed to elicit high titres of antibody against these antigens. These vaccines are useful in livestock farming to help control *scours*; diarrhoea caused by *E. coli* is one of the major causes of death of piglets. The cost of the losses caused by diarrhoea in piglets in Europe during 1975 was estimated at more then £100m. Protection against enterotoxigenic *E. coli* can be achieved by vaccinating sows with a preparation containing K88 antigen. Newborn piglets have no immunoglobulin, but receive antibodies, including those against K88$^+$ *E. coli,* from the vaccinated sows. Colostrum containing antibodies is absorbed by the piglets during the first 24 h after birth. Subsequently, antibodies in milk are retained in the alimentary tract of the piglets and prevent colonization by K88$^+$ *E. coli.* Some protection against K88$^+$ *E. coli* can also be obtained by breeding strains of livestock which cannot be colonized effectively by these bacteria.

Species of *Shigella* which cause dysentery (*S. sonnei, S. flexneri* and *S. dysenteriae*) all have virulence plasmids. The plasmids have at least two virulence determinants; they encode the production of specific O-antigens in the cell wall and they also enable bacteria to invade epithelial cells in the wall of the large intestine. There are also many chromosomal virulence determinants. Strains of *E. coli* causing a dysentery-like disease have a plasmid very similar to that found in *Shigella.*

b Generalized infections caused by *E. coli* *E. coli* strains can cause generalized infections—they invade the body of the animal and are found in large numbers in the blood and various organs. In severe cases, death results from a general septicaemia. Strains of *E. coli* responsible for generalized infections are an important cause of disease and death among calves and chicks. Particular serotypes are usually responsible for these infections and these usually harbour plasmids which enhance their virulence. This was discovered by Smith (1976) who found that most of the strains causing generalized infections in livestock had a

ColV plasmid, that is, a plasmid coding for the synthesis of colicin V. The ColV plasmids increased bacterial resistance to host-defence mechanisms and produced an approximately 100-fold reduction in the LD_{50} (50% lethal dose) of *E. coli* for chicks. It was subsequently found that neither colicin V nor immunity to colicin V was responsible for these effects. Two virulence determinants have been found on ColV plasmids. One of these increases bacterial resistance to antibacterial defence mechanisms which are mediated by the complement proteins present in blood. The other enables bacteria to synthesize a hydroxamate so that they can accumulate iron more effectively from their surroundings. There is almost no free iron in blood, lymph or tissue fluids; it is assosiated with transferrin and other proteins which bind iron very tightly. Growth of bacteria in these fluids therefore depends on their ability to produce iron-chelating agents such as hydroxamate which remove iron from transferrin so that it can be transported into the bacterial cells. An iron-uptake system of this type is specified by many ColV plasmids present in strains causing generalized infections. The hydroxmate determinant on ColV plasmids is associated with IS*1* sequences. A similar determinant associated with IS*1* sequences was found on plasmids from *Salmonella* spp. which were frequently responsible for septicaemia and meningitis. Several other virulence plasmids have been found in *Salmonella* spp. but their functions are unknown. A plasmid present in the fish pathogen *Vibrio anguillarum* also codes for an iron-binding compound and increases the virulence of the bacterium.

Pathogenic species of *Yersinia, Y. pestis, Y. pseudotuberculosis and Y. enterocolitica,* all have a virulence plasmid. *Yersinia pestis* is the cause of bubonic and pneumonic plague. The virulence plasmids in these bacteria encode outer membrane proteins.

Other toxins Certain strains of *Staphylococcus aureus* produce exfoliative toxin and can cause scalded-skin syndrome in susceptible people, mainly in young children. The skin becomes loosened and may eventually peel off in large sheets. Although the infection is initially confined to the skin around the mouth and nose, it can rapidly spread to most parts of the body. Production of exfoliative toxin by several *S.aureus* strains depends on the presence of a plasmid which presumably encodes the protein. Exfoliative toxin is specified by chromosomal genes in some strains.

The gene for tetanus neurotoxin produced by *Clostridium tetani* is on a plasmid. Plasmids also encode the toxin made by *Bacillus anthracis,* the cause of anthrax, and the protein crystals made by *Bacillus thuriengiensis* var. *israeliensis* which are toxic to the larvae of certain dipteran insects.

Crown gall This is a cancer of plants caused by the bacterium *Agrobacterium tumefaciens.* The particularly interesting feature of the disease is that it is induced by part of a plasmid which is transferred from bacteria to plant cells causing them to grow to form tumours. The disease is especially important in parts of California and in Australia where it can greatly reduce yields from fruit trees if it is not controlled. *A. tumefaciens* can cause tumours in many dicotyledonous plants, but the disease has not yet been established in a monocotyledonous plant. Infections can be established only in freshly wounded plants.

Plant cells from tumours can be grown in culture and, unlike cultures derived from normal plant cells, they can be grown without added plant hormones. Furthermore, once a tumour has been induced by *A. tumefaciens,* the plant cells continue to grow in the absence of bacteria. Tumours which are transplanted to healthy plants also continue to grow.

A. tumefaciens is a Gram-negative rod-shaped bacterium belonging to the *Rhizobiaceae*. This family also includes species of *Rhizobium* which form nodules on leguminous plants and have plasmids which enable them to fix nitrogen. The first indication that plasmids were essential for tumour formation came from analyses of variants of *A. tumefaciens* which had lost the ability to cause Crown Gall. Some strains gave rise to these variants at high frequencey if they were grown at temperatures above the optimum for growth. Analyses of DNA from avirulent variants by dye-buoyant density centrifugation showed that they had lost a plasmid. The variants regained virulence when they received a tumour-inducing (Ti) plasmid from another strain of *A. tumefaciens*. Ti plasmids are very large, having molecular weights of between 90×10^6 and 160×10^6, and can be transferred between strains by conjugation.

Only a small fragment of the Ti plasmid is stably maintained by the plant cells which form tumours. The fragment (called T-DNA) has a molecular weight of about 15×10^6 and in some tumours is maintained at about 20 copies per plant cell. The particular region of Ti plasmids which is maintained by plant cells was identified by generating fragments of plasmids with restriction endonucleases and determining which of these would hybridize (that is, would anneal by complementary base-pairing) with DNA derived from plant tumours. RNA transcripts of the plasmid fragment have also been detected in tumours.

The precise mechanism of tumour induction by the Ti plasmid is unknown. The T-DNA in Crown Gall tumours is integrated at various sites in the plant chromosomes. T-DNA causes changes in the concentrations of plant hormones; it increases the concentration of indoleacetic acid and cytokinin. Genes encoding enzymes involved in the production of these two plant hormones were found on the Ti plasmid. *Pseudomonas syringae* pr. *savastonoi,* another plant pathogen, also has plasmid genes which encode the synthesis of indoleacetic acid.

In addition to stimulating the growth of plant cells, Ti plasmids also cause them to synthesize compounds known as opines. Opines are not produced by normal plant tissue and have not been found elsewhere in nature other than in Crown Gall. The first two opines to be discovered were octapine and nopaline, which are derivatives of arginine. A particular Ti plasmid induces plant cells to synthesize either octapine or nopaline; only one plasmid has been found to induce the synthesis of both these opines. Opine synthases are presumably encoded by genes on the T-DNA which is maintained by plant cells. The significance of opines in the disease became clearer when it was found that plasmids which stimulated octapine production by tumerous plant tissue also encoded enzymes which enabled their host bacterium to use octapine as a source of carbon and nitrogen. Similarly, Ti plasmids which induced plant cells to synthesize nopaline enabled bacteria to use this particular opine as a growth substrate. The bacterial enzymes for opine catabolism convert octapine to arginine and pyruvate, and nopaline to arginine and α-ketoglutarate. Ti plasmids also enable bacteria to accumulate the corresponding opines more efficiently. Presumably, the catabolic enzymes and permeases are encoded by plasmid genes.

Bacteria containing Ti plasmids therefore seem to have evolved a most ingenious method for tapping the synthetic abilities of plants to provide them with exclusive sources of carbon and nitrogen. The ability to utilize opines is an uncommon trait amongst microorganisims so it is unlikely that bacteria other than those containing the appropriate Ti plasmid would be able to benefit from the opines produced by plant tumours.

The ability of the Ti plasmid to transfer and to maintain part of itself in plant cells has opened the way for inserting other genes into plants. These can be attached to the appropriate part of the Ti plasmid, so that they can be transferred from *A. tumefaciens* to plant cells as part of the Ti plasmid. Plasmid vectors have been made containing T-DNA from which the tumour-inducing genes have been removed. Using these vectors, several bacterial drug-resistance genes (chloramphenicol acetyl transferase, neomycin phosphotransferase and dihydrofolate reductase) have been expressed in plant cells. The promoter of the nopaline synthase gene can be used to express these foreign genes.

Bacteriocins

Bacteriocins are antibacterial proteins produced by bacteria. Most bacteriocins have a narrow spectrum of action and are lethal only for bacteria which are closely related to strains which produce them. Colicins are bacteriocins produced by *E. coli* and closely related members of the *Enterobacteriaceae* such as *Shigella sonnei;* they are bactericidal for many enterobacteria. About 40% of *E. coli* strains from man or animals are colicinogenic (Col$^+$).

Substances analogous to colicins are produced by many Gram-positive and Gram-negative genera. For example, some streptococci produce streptococcins, and Staphylococcins are produced by certain strains of *Staphylococcus.* In each case, the active component of the bacteriocin is protein.

Colicins are specified by Col plasmids. There are no reports of natural isolates which have chromosomal genes coding for colicins although such strains can be made in the laboratory. Each Col plasmid also confers immunity to the particular type of colicin it encodes, so that a strain specifying a type A colicin will be immune to colicin A but may be sensitive to many other types. It is less clear whether other types of bacteriocin besides colicins are usually plasmid-encoded since fewer have been studied, but several staphylococcins and streptococcins and a perfringocin (produced by *Clostridium perfringens*) are known to be plasmid-encoded. The rest of this section is mainly about colicins which have been the subject of most investigations on bacteriocinongeny.

Classification Colicins (and the Col plasmids which code for them) are given names such as 'A-CA31' which indicates the colicin type, 'A' in this case, and the strain which was first shown to produce this particular colicin, *E. coli* strain CA31. When the plasmid ColA-CA31 is transferred to another strain it retains the same name.

Colicins were classified into about 20 different types by Fredericq on the basis of their action on various colicin-resistant strains. Fredericq's classification has provided the basis for all subsequent studies on colicins. Some of the resistant strains used by Fredericq were natural isolates sensitive to only a few colicin types, and others were colicin-resistant mutants derived from sensitive strains. Colicins of a particular type are defined as all those colicins which are ineffective against a particular colicin-resistant strain. In practice, classifying colicins in this way is complicated by the fact that many colicin-insensitive strains are resistant to more than one type of colicin. In addition, bacteria often produce several types of colicin and an individual Col plasmid can code for two or more kinds of colicin. (Unfortunately, a set of strains which are uniquely sensitive to particular types of colicin does not exist. A set of this type would make it much easier to identify the colicins made by isolates of *E. coli.*) Sub-divisions of types are based on colicin immunity. Colicins E2 and E3 are indistinguishable on the basis of their action on colicin-resistant mutants, but bacteria which have a ColE2 plasmid are immune only to other E2 colicins and not to E3 colicins. Similarly, ColIa$^+$ bacteria are immune to Ia colicins but not to Ib colicins.

Structure of bacteriocins Bacteria produce such a wide variety of inhibitors that it is sometimes difficult to decide how many of these should be classified as bacteriocins. Inhibitors which have proved on closer examination to be acids or hydrogen peroxide can clearly be excluded. A substance is usually called a bacteriocin if its active component is protein and if it is effective only against bacteria related to the producing strain. These criteria are usually easily applicable to substances produced by Gram-negative strains; these usually have a narrow spectrum of action which can be easily demonstrated. In addition, they are more easily purified (their synthesis can often be induced by agents such as mitomycin C) and they are usually simple proteins which do not have extensive amounts of bound carbohydrate or lipid. It is usually easy to demonstrate whether the bacteriocin is plasmid-specified. As more and more of the substances which are judged to be bacteriocins by other criteria prove to be plasmid-specified, it seems that this characteristic might also be included in any definition of these substances.

Substances produced by Gram-positive bacteria are sometimes more difficult to classify. They are often active against many unrelated groups of Gram-positive bacteria and are more difficult to purify. Although it is clear that some are plasmid-specified, fewer genetic studies have been made and it is less clear whether this is a common feature of bacteriocins produced by Gram-positive bacteria.

Inhibitory substances which appear to be bacteriocins because they are inactivated by proteases and because they act only on a few related bacteria sometimes prove to be particles resembling bacteriophage tails when they are examined by electron microscopy. Structures of this type produced by strains of *Pseudomonas aeruginosa* have been studied most extensively. They are called aeruginocins or pyocins (*P. aeruginosa* was formerly called *P. pyocyanea*) but they are clearly distinct from the low molecular weight bacteriocins and are often called *defective bacteriophages* rather than bacteriocins. They are specified by chromosomal genes and have immunological cross reactions with functional temperate bacteriophages found in *Pseudomonas*.

The molecular weights of colicins and other bacteriocins range from about 12 000 to 90 000. At least some colicins are released from bacteria in the form of a complex of two proteins. The larger of the two-proteins produces the lesion which eventually kills sensitive cells. The smaller sub-unit is the colicin immunity protein.

Although it inhibits the action of colicin *in vitro* it does not prevent the larger sub-unit from killing sensitive cells. Indeed, the larger sub-unit by itself is much less less effective at killing cells than the colicin-immunity protein complex. The immunity protein binds tightly to the colicin.

Several colicins have similar sequences, indicating that they have evolved from common ancestors. The similarity is often confined to a particular region of the molecules, as discussed in the next section.

Several of the bacteriocins produced by Gram-positive bacteria (streptococcins, staphylococcins, and lactocins produced by *Lactobacillus*) have molecular weights of 10 000 to 25 000 and are proteins associated with lipid and carbohydrate. These bacteriocins tend to aggregate into larger complexes.

Lethal effects of bacteriocins Colicins are much larger molecules than antibiotics such as streptomycin or chloramphenicol and the cell membranes of bacteria are, of course, impermeable to such large proteins. The question of how large proteins can kill bacteria is particularly intriguing because it is clear from biochemical studies that colicins act by a variety of different mechanisms, and some of them kill cells by affecting intracellular targets such as DNA or ribosomes. It was thought that colicins might attack intracellular targets from outside the cell by activating membrane-bound enzymes, but recent studies have shown that parts of some colicins are indeed able to penetrate the cell envelope of *E. coli* (despite their large size) and act as DNases or RNases. Other colicins kill cells by increasing the permeability of the cytoplasmic membrane. Colicins of this type probably penetrate only as far as the cytoplasmic membrane.

Although colicins can kill in a number of different ways, they all initially bind to receptors on the bacterial surface. Colicin receptors are proteins in the outer membrane. The outer membrane of *E. coli* comprises a bilayer of phospholipid molecules together with proteins and lipopolysaccharide.

Many colicin receptors also function as receptors for metabolites which are transported into the *E. coli* cell. The receptors for colicins B, E, K and M are essential for transporting ferri-enterochelin, vitamin B_{12}, nucleosides and ferrichrome, respectively. Enterochelin and ferrichrome are iron-chelating agents which enable bacteria to accumulate iron. Other proteins in the cell envelope, in addition to receptors, are essential for the transport of metabolites and for the action of several colicins. One of these is the *tonB* protein. *TonB*$^-$ mutants are defective in an envelope protein which is essential for transporting ferri-enterochelin and other metabolites through the cell wall. These mutants are also resistant to several types of colicin, indicating that the *tonB* protein also transports colicins (or parts of them) through the wall.

Although colicin receptors are essential for the action of colicins on ordinary cells, they may be unnecessary if parts of the cell wall are removed or disrupted to allow colicins to reach the inner parts of the cell envelope. Sphaeroplasts made by disrupting the walls of cells which do not have functional colicin receptors can be sensitive to colicins although the intact cells are completely colicin-resistant.

Most types of colicin (including colicins, A, E1, Ia, Ib and K) are bactericidal because they cause the formation of channels in the cytoplasmic membrane which then becomes permeable to ions. Important cations, such as K^+, are released from colicin-treated cells. This dissipates the potential difference which exists across the membrane and thereby inhibits all those reactions which depend on the energized state of the membrane. These reactions include the transport of many sugars and amino acids and the movement of bacterial flagella. The loss of membrane potential also leads to a decrease in the intracellular concentration of ATP since this is used by colicin-treated cells in an attempt to maintain the energized state of the membrane which is dissipated by the colicin. Consequently, colicins such as E1 and K inhibit all ATP-dependent reactions, including the synthesis of proteins and nucleic acids. In addition, loss of cations from cells inhibits all enzymes which require them as co-factors.

The amino acid sequence of the C-terminal part of colicins A, E1, I and K are partially similar. This part of the molecule appears to be involved in the interaction with the membrane. The N-terminal parts of these colicins bind to different receptors and are not homologous.

At least one type of staphylococcin acts like colicins such as E1 and K. A bacteriocin produced by *Clostridium butyricum* also inhibits energy-dependent reactions in sensitive cells, but in this case the bacteriocin directly inhibits the membrane-bound ATPase of sensitive bacteria.

Colicins E2, E3 and cloacin DF13 (a bacteriocin produced by *Enterobacter cloacae* DF13 which is closely related to colicin E3) do not kill sensitive cells primarily through their effects on the energized state of the cytoplasmic membrane. DNA is degraded in colicin E2-treated cells, and the 16S rRNA molecule is cleaved 50 nucleotides from the 3′ end in cells treated with colicin E3 or with cloacin DF13. Purified preparations of these colicins can act as nucleases *in vitro* so it seems that these colicins (or at least parts of them) penetrate the cell envelope to attack their intracellular targets. The first attempts to show that these bacteriocins had nuclease activity *in vitro* were unsuccessful because the preparations used were in fact complexes of colicin molecules bound to immunity protein. Colicins E2 and E3 have molecular weights of about 50 000 and the immunity proteins have molecular weights of about 10 000. They are released in the form of a complex; each colicin molecule has a molecule of immunity protein firmly attached to it. The immunity protein inhibits the *in vitro* action of the colicins but does not inhibit the action of the colicins on sensitive cells. Indeed, the complex of colicin + immunity protein is much more effective against sensitive bacteria than the colicin alone. The immunity protein should perhaps be seen as an intrinsic part of the active colicin than as simply a molecule which confers immunity on Col$^+$ bacteria. The immunity protein specified by ColIa is found in the cytoplasmic membrane. Colicin Ia kills cells by causing membrane depolarisation.

Analyses of several colicins (and of fragments of them) indicates that it is the central portion of these molecules which binds to all receptors. The N-terminal parts of colicins E1 and E3, which are rich in glycine residues, are involved in the transport of these molecules across the cytoplasmic membrane after they have attached to receptors. The C-terminal parts of colicins appear to have the catalytic activity which eventually kills sensitive cells.

Colicins E1, A and Ib, which all cause depolarisation of the cytoplasmic membrane, have a hydrophobic C-terminal domain apparently essential for forming channels in the membrane. Colicin E2 closely resembles colicin E3 except in the C-terminal part; colicin E3 is an RNAase, whereas colicin E2 is a DNAase. The immunity proteins also bind to the C-terminal parts.

It has been suggested that colicins might be split into C-terminal and N-terminal fragments at the cell surface. The N-terminal part of colicins E2 and E3, for example, might be left outside the cytoplasmic membrane, while the C-terminal parts enter the cell and act as nucleases. The immunity protein is presumably separated from the C-terminal fragment before it enters the cell. In fact, there is no evidence either for or against this view. Several bacterial toxins, including cholera toxin and diphtheria toxin, are split into two fragments after they have adsorbed to receptors on mammalian cells.

The points at which removal of immunity protein occur during the passage of colicins through the cell envelope is not yet known. The N-terminal region is in fact hydrophobic which could facilitate transport of the colicin across the cell envelope. The C-terminal fragments of some colicins are a highly basic proteins. They may be neutralized by the acidic immunity protein which may also facilitate transport across membranes, either during their release from Col$^+$ cells or during their action against sensitive cells.

Two bacteriocins, colicin M and a pesticin (produced by *Yersinia pestis*), kill cells because they weaken the cell wall, so that the cells lyse if they are kept in hypotonic medium. Pesticin catalyses hydrolysis of the β-1,4 bond between N-acetylglucosamine and N-acetylmuramic acid to destroy the integrity of the mucopeptide backbone.

Col plasmids Plasmids specifying colicins fall into two groups: Group I Col plasmids are nonconjugative plasmids which have molecular weights of about 5×10^6. ColE1-K30 and CloDF13 are examples of this class. Group II Col plasmids are conjugative with molecular weights of 50 to 85 $\times 10^6$; some code for more than one type of colicin and they may also have genes for drug-resistance or for virulence. Most plasmids coding for colicins V or B are conjugative and code for F-like pili. Plasmids specifying colicin I usually specify I pili. About 40% of I-like R plasmids code for colicin I synthesis. ColV, I-K94 is an exceptional F-like plasmid which codes for colicin I as well as colicin V. Almost all the *tra* regions of ColV,I-K94 and F are homologous as judged by the heteroduplex technique. This is in agreement with the results of genetic studies which show that most of the *tra* genes of the F plasmid can be complemented by the corresponding genes of ColV,I-K94.

Less is known about the plasmids coding for other types of bacteriocin. Only three bacteriocinogenic plasmids from Gram-positive bacteria have been characterized. These code for the synthesis of a staphylococcin, a streptococcin and a perfringocin (produced by *Clostridium perfringens*).

Bacteriocin synthesis Usually only about 0.1% of cells in a broth culture of Col$^+$ bacteria produce colicin. These few cells release colicin for several hours. Synthesis of some, but not all types of colicins, leads to death of the producing cell.

Colicin synthesis is largely controlled by a protein repressor encoded by the chromosomal *lexA* gene. The repressor binds to the operators of the genes for colicins. When the bacterial DNA is damaged (for example by ultraviolet irradiation) a protease specified by the bacterial *recA* gene is activated and cleaves the lexA protein to allow induction of colicin synthesis. DNA damage activates the so-called 'SOS-response' which leads to the expression of several genes, including those involved in DNA repair. The activated genes all have an

'SOS-box'—a specific operator sequence which binds the *lexA* repressor. Colicin genes also have an SOS-box associated with them.

Colicins and temperate phages such as lambda are induced by similar agents and in both cases induction is prevented by the *recA* mutation. In the case of lambda, the phage itself encodes a repressor (the product of the cl gene) which is sensitive to the *recA* gene product.

Colicin induction also leads to the synthesis of a lysis protein which releases colicin from cells. Lysis proteins encoded by several Col plasmids have molecular weights of about 4000 and are found in the cell envelope. After induction, colicin accumulates inside the cell for 1–2 hours until the lysis protein makes the membrane sufficiently permeable to allow colicin release. Several other cell proteins are also released, along with smaller molecules, and the cells lose viability. On some Col plasmids, the genes for colicin, colicin immunity protein and lysis protein are all transcribed from the same promoter so that induction stimulates production of all three.

Microcins About 10% of *E. coli* strains (and of several other kinds of enterobacteria) produce another kind of antibiotic, a microcin, which can be distinguished from the classical colicins by a number of criteria. They are amino acid derivatives or peptides of low molecular weight (250–5,000), they are not inducible (by mitomycin C, for example), their synthesis is not lethal for the producing cell and it is not accompanied by cell lysis. Several microcins are plasmid-encoded.

Microcin 15m (molecular weight 240) is a methionine analogue which acts by inhibiting homoserine-*o*-trans succinylase, an enzyme which regulates methionine metabolism in *E. coli.* Microcin 140 is a peptide having a molecular weight of less than 1000, which acts similarly to colicin E1. Microcin 17 is a peptide (molecular weight 4000) which inhibits DNA synthesis.

Ecology of bacteriocins Bacteriocins are commonly produced by bacteria so it is assumed that they confer a selective advantage by killing other bacteria which are in competition with the bacteriocinogenic strains. It could be argued that the narrow spectrum of action of many bacteriocins is an example of Darwin's proposal that competition between closely related species will be particularly intense because these species, which share many common features in terms of nutrition and habitat, will inevitably be competing in a similar environment. Antibiotics produced by *E. coli* which enable it to inhibit the growth of, say, *Bacillus subtilis* are unlikely to be selected because *E. coli* is unlikely to find itself in the same environment as *Bacillus subtilis;* any such antibiotic is unlikely to confer a selective advantage. On the other hand, in an ecological niche which can support only a limited number of *E. coli* bacteria, a Col^+ strain might be at a selective advantage if it can remove its closely related rivals.

The molecular basis for the narrow spectrum of bacteriocins such as colicins appears to be a consequence of the need for specific cell wall proteins which apparently transport colicins to the cytoplasmic membrane, and are perhaps also involved in processing colicins by cleaving them into two fragments. Purified colicin E3 can act as a ribonuclease against the ribosome of *Bacillus stearothermophilus,* which is completely unrelated to members of the *Enterobacteriaceae.* Cells of *B. stearothermophilus* are colicin E3-resistant presumably because the colicins cannot penetrate cells which lack the appropriate receptors.

The extent to which colicins confer a selective advantage to cells rather than simply to plasmids is unclear. Conceivably, Col plasmids might be parasites, rather like bacteriophages, which confer little selective advantage on bacterial cells. Genes coding for colicins have apparently evolved an excellent method for securing their survival. Even though they may render a host cell uncompetitive in some ways, it cannot dispense with the Col plasmids because it may then become susceptible to remaining Col^+ cells. It has also been pointed out that colicin receptors are almost invariably involved in transporting useful metabolites into the cell. Populations of bacteria may not therefore become colicin-resistant without incurring the penalty of decreased efficiency in the uptake of metabolites.

The theoretical advantages of colicinogeny seem obvious enough, but there is little evidence upon which to assess how important or effective colicins or other bacteriocins are in natural environments. One of the difficulties in evaluating experiments which compare the survival of Col^+ and Col^- strains in the alimentary tract is that some Col plasmids code for other characteristics, beside colicins, which increase survival. For example, many ColV plasmids code for an iron-uptake system and also confer resistance to the bactericidal effects of complement. Both of these factors could confer a selective advantage on strains in the alimentary tract.

Experiments designed to detect a selective advantage conferred by bacteriocinogeny in *Bacteroides* spp have also been unsuccessful. Members of the *Bacteroides* genus are obligate anaerobes which are major components of the intestinal flora (unlike *E.coli* which comprises less than 1% of the bacterial cells in the alimentary tract). Bacteriocin-producing strains form the minority of the total *Bacteroides* population of most people. Bacteriocin-sensitive strains co-exist with bacteriocinogenic strains for many weeks even though the strains are highly sensitive to bacteriocins *in vitro*. Recent experiments indicate that bacteriocinogeny might be a significant factor in the establishment of strains of *Streptococcus mutans* (the bacterium largely responsible for dental caries) in dental plaque. Strains of *Agrobacterium radiobacter* which produce agrocin 84 (a nucleotide derivative) are used extensively to control Crown Gall disease. Attempts are also being made to use preparations of the agrocin itself. An interesting feature of the biological control of Crown Gall is that sensitivity to agrocin 84 is conferred by a gene on the nopaline type of Ti plasmid found in virulent strains of *A. tumefaciens*. Strains which become agrocin-resistant through loss of the Ti plasmid are therefore avirulent.

Metabolic plasmids

The genus *Pseudomonas* includes bacteria which can grow on a very wide range of organic compounds. They play an important role in recycling carbon in the biosphere by degrading complex organic compounds to simpler forms which can be used as sources of carbon and energy by living organisms. Many of the degradative enzymes of pseudomonads are specified by chromosomal genes, but it has recently become clear that plasmid-encoded enzymes are often involved.

Degradative plasmids Degradative plasmids enable *Pseudomonas putida* and related bacteria to grow on substrates such as toluene, xylene, octane, camphor, naphthalene, salicylate and nicotinic acid (Table 7). Plasmid-encoded enzymes convert these growth substrates to metabolites such as acetaldehyde, pyruvate and acetate which can then enter metabolic pathways catalysed by chromosomal enzymes. Most degradative plasmids code for at least 10 enzymes which are involved in the catabolism of a particular substrate,

The SAL, NAH, ASL and TOL plasmids convert salicylate, naphthalene, alkyl benzene sulphonates and toluene, respectively, to catechols which are then degraded further by plasmid-encoded enzymes to acetaldehyde and pyruvate. The pathway for catechol catabolism used here differs from the pathway which is specified by chromosomal enzymes in *P. putida,* in which intra-diol cleavage of catechol occurs between the *ortho* hydroxyls. This pathway is not used when a TOL or similar plasmid is present in *P. putida* because induction of the catechol 1,2-oxygenase requires intracellular accumulation of catechol, which does not occur if a TOL plasmid is present. Enzymes coded for by the SAL, NAH, TOL and ASL plasmids catabolize catechols via a pathway which involves extradiol or *meta*-cleavage by a catechol 2,3-oxygenase. As this is substrate-induced by the appropriate growth substrates, the catechols never accumulate intracellularly to a sufficient extent to induce the *ortho* pathway.

Degradative plasmids enable bacteria to grow on synthetic compounds and may therefore make an important contribution to the removal of substances which can act as pollutants. Plasmids enable bacteria to degrade the widely used herbicide 2, 4-D (2, 4-dichlorophenoxyacetic acid), chlorinated hydrocarbons such as *p*-chlorbiphenyl, and alkyl benzene sulphonates (detergents). Attempts are being made in several laboratories to use the metabolic potential of degradative plasmids by constructing strains which can be used for pollution control or for chemical syntheses.

Degradative plasmids are usually conjugative and have high molecular weights, ranging from about 50×10^6. to about 200×10^6. They belong to various P incompatibility groups and are usually compatible with each other. Mitomycin C is a particularly effective curing agent for degradative plasmids. The TOL, SAL, OCT and NIC plasmids can each dissociate into two or more separate plasmids which are capable of independent replication. Dissociation gives rise to a non-conjugative plasmid, which specifies the degradative pathway, and to a separate conjugative plasmid. The nonconjugative plasmids derived from TOL and SAL can become inserted into apparently unrelated plasmids. (The fragment of TOL which can be inserted into unrelated DNA molecules has a molecular weight of about 40×10^6.) TOL plasmids can also dissociate to lose a fragment which has a molecular weight of about 30×10^6. This appears to involve a recombination at directly repeated 1,400 base pair sequences, and resembles the dissociation of certain R plasmids at directly repeated IS*1* sequences.

Some degradative plasmids, such as TOL and NAH, have a wide host-range and can be transferred by conjugation to *E.coli* as well as to pseudomonads. TOL$^+$ *E. coli* strains are able to use toluene and xylene as sole carbon sources but they grow more slowly than TOL$^+$ strains of *Pseudomonas,* partly because the plasmid genes are not expressed efficiently in *E. coli.*

Table 7 Degradative plasmids in *Pseudomonas*

Plasmid	Primary substrate	End-product of plasmid-encoded pathway
TOL	toluene, *m*-xylene, *p*-xylene	→ Salicylate → catechol or alkylcatechol → acetaldehyde or higher alkanal + pyruvate
SAL	salicylate	→ Salicylate → catechol or alkylcatechol → acetaldehyde or higher alkanal + pyruvate
NAH	naphthalene	→ Salicylate → catechol or alkylcatechol → acetaldehyde or higher alkanal + pyruvate
ASL	alkyl benzene sulphonates (R, SO_3^-)	→ catechol or alkylcatechol → acetaldehyde or higher alkanal + pyruvate
OCT	octane $CH_3(CH_2)_6CH_3$	octanol and octanal
CAM	camphor (CH_3, O, $C(CH_3)_2$)	isobutyrate
NIC	nicotinic acid (N, COOH)	maleate and fumarate

(Partly derived from Williams, P. A. (1978). 'The Biology of Plasmids.' In *Companion to Microbiology* pp. 77–108. Edited by A. T. Bull and P. M. Meadow; Longman.)

Other metabolic plasmids Metabolic plasmids sometimes become apparent because they confer the ability to ferment unusual substrates so that strains give unexpected results when being identified or classified. Strains of *Salmonella* are almost always Lac^-, that is, they are unable to ferment lactose. Lac^+. strains of *Salmonella typhi* and other salmonellae have been isolated on rare occasions and have been found to harbour plasmids which confer the Lac^+ phenotype. Lac^+ plasmids have also been found in strains of *Serratia, Streptococcus lactis* and *Proteus.* It is also interesting that the lactose operon was found to be encoded by a transposon, Tn*951, in Yersinia enterocolitica.* Plasmids can also confer the unusual ability to ferment sucrose on strains of *Salmonella.* Almost all plasmids coding for K88 antigen, which increases the virulence of enterotoxigenic *E. coli,* also confer the ability to ferment raffinose. Some raf^+ plasmids also specify H_2S production. Genes for raffinose utilization are flanked by direct repeats of IS*1* and are transposable. Genes for K88 antigen or H_2S production form part of the transposon when they are linked to the raf^+ genes.

Metabolic plasmids can occasionally lead to misidentification or to delays during bacteriological diagnosis of infectious diseases. Plasmids are also relevant to diagnostic bacteriology in other ways. Phage typing of *Salmonella typhimurium* is an important tool in epidemiological investigations of diseases caused by this bacterium. The phage type of a strain is sometimes determined by a plasmid or prophage which restricts the growth of the phages used for typing. The loss or gain of a plasmid can change the phage type. An example of such a plasmid is the Δ plasmid which determines the phage type of *S. typhimurium* type 29 (p.56).

An important group of metabolic plasmids discovered recently are those which enable strains of *Rhizobium* to nodulate and fix nitrogen in the roots of leguminous plants, such as peas, beans and clover. The various species of *Rhizobium* form a symbiotic relationship only with particular legumes; for example, *R. leguminosarum* nodulates peas, whereas *R. phaseoli* nodulates beans. Both the specificity of nodulation and at least some of the genes for nitrogen fixation are specified by plasmids. Strains of *R. leguminosarum* which have lost the ability to nodulate peas, regain this ability when they receive a conjugative plasmid from a nodulating strain. Furthermore, a nodulating strain of *R. leguminosarum* can transfer the ability to nodulate peas to *R. phaseoli.* Some *R. phaseoli* recipients can no longer nodulate beans because they have lost the plasmid which confers this characteristic. Plasmids in *R. leguminosarum, R. trifolii* and *R. meliloti* also have at least some of the genes for nitrogen fixation (*nif* genes), including the genes for nitrogenase. The molecular weights of these large plasmids are greater than 90×10^6. The presence of *nif* genes on *Rhizobium* plasmids was shown by hybridizing plasmid fragments (generated by restriction endonucleases) with cloned fragments of the *Klebsiella pneumoniae* chromosome which encode nitrogenase and other proteins essential for nitrogen fixation.

Plasmids are involved in antibiotic production in several species of *Streptomyces.* Plasmid SCPI found in *Streptomyces coelicolor* has structural genes for the synthesis of the antibiotic methylenomycin A. Plasmid genes are also involved in the synthesis of other antibiotics including chloramphenicol and kasugamycin. In many cases, it appears that the structural genes for the enzymes for antibiotic production are chromosomal and that plasmid genes can modify their expression.

Summary

E. coli is an important cause of diarrhoea in both man and animals. Enterotoxigenic strains produce either a heat-labile toxin (LT), a heat-stable toxin (ST), or both, which are usually specified by conjugative plasmids. LT is related to the toxin produced by *Vibrio cholerae* and comprises two sub-units with molecular weights of 25 000 and 11 500. ST is a peptide with a molecular weight of about 5000. A transposon specifying ST has been discovered. The virulence of enterotoxigenic strains is increased by plasmids coding for pili which enable bacteria to stick to the wall of the small intestine.

Most of the strains causing generalized infections in livestock have ColV plasmids which enhance their pathogenicity. Two virulence determinants have been found on these plasmids: genes for resistance to complement and genes for iron accumulation. *Shigella* strains causing dysentery have virulence plasmids and plasmid genes contribute to the invasiveness of *Yersinia pestis*. The exfoliative toxin of some *S. aureus* strains, and the toxins made by *Clostridium tetani* and *Bacillus anthracis* are plasmid-specified.

The Ti plasmid of *Agrobacterium tumefaciens* enables this bacterium to cause Crown Gall in dicotyledonous plants. A fragment of the plasmid is maintained by plant cells which form a tumour. The plasmid fragment induces tumour cells to form opines; some opines are unusual derivatives of basic amino acids. Ti plasmids enable *A. tumefaciens* to use opines as sources of carbon and nitrogen. They also encode plant hormones.

Bacteriocins are antibacterial proteins produced by bacteria. Colicins are bacteriocins produced by *E. coli* and other enterobacteria which have a spectrum of action confined to the *Enterobacteriaceae.* They are specified by either Group I Col plasmids (small, nonconjugative) or Group II Col plasmids (large, conjugative) and are classified into about 20 types according to their effects on a set of colicin-resistant strains. Col plasmids confer immunity to the colicins they encode. Some colicins, including colicins E2 and E3, are released from Col$^+$ cells as a complex of two proteins. The smaller sub-unit is an immunity protein which inhibits the *in vitro* action of colicin but does not prevent it from killing sensitive cells. Colicins bind to receptors in the outer membranes of bacteria. Some types of colicin then reach the cytoplasmic membrane where they cause ion-permeable channels to form. Intra-cellular cations are released so that the potential difference across the membrane is dissipated. The ATP concentration in the cell then decreases, inhibiting all energy-dependent reactions. Colicin E2 is a DNase and colicin E3 is an RNase which specifically attacks the 16S RNA of ribosomes. Parts of these colicins penetrate sensitive cells.

Bacteriocins produced by staphylococci, streptococci and lactobacilli are proteins associated with lipid and carbohydrate. Some are plasmid-specified.

Col$^+$ strains often persist longer in the alimentary tract than Col$^-$ strains, but this might not result from the bactericidal effects of colicins since some Col plasmids specify other features which enhance survival. Bacteriocinogenic strains are often found to coexist with bacteriocin sensitive strains for long periods in natural environments.

Degradative plasmids enable *Pseudomonas putida* and related pseudomonads to grow on substrates such as toluenes, xylene, octane, camphor, naphthalene, saclicylate, alkyl benzene sulphonates and nicotinic acid. Plasmids also enable bacteria to grow on synthetic compounds such as the herbicide 2, 4-D. Many degradative plasmids encode 10 or more inducible enzymes of catabolic pathways. They are large (molecular weights range from 50×10^6 to 200×10^6) conjugative plasmids which belong to various P incompatibility groups. Some have a wide host range.

Metabolic plasmids found in enterobacteria code for the ability to ferment sugars such as lactose, sucrose and raffinose. Plasmids in *Rhizobium* were recently shown to be essential for nodulation of legumes and for nitrogen fixation.

References

ELWELL, L. P. and SHIPLEY, P. L. (1980). Plasmid-mediated factors associated with virulence of bacteria to animals. *Annual Review of Microbiology* 34:465–496.

HOPWOOD, D. A. (1978). Extrachromosomally determined antibiotic production. *Annual Review of Microbiology* 32: 373–392.

KONISKY, J. (1982) Colicins and other bacteriocins with established modes of action. *Annual Review of Microbiology* 36: 125–144.

NUTI, M. P., LEPIDI, A. A., PRAKASH, R. K., SCHILPEROORT, R. A. and CANNON, F. C. (1979). Evidence for nitrogen fixation (*nif*) genes on indigenous *Rhizobium* plasmids. *Nature* 282: 533–5.

SMITH, H. WILLIAMS (1976). Neonatal *Escherichia coli* infections in domestic animals: transmissibility of pathogenic characteristics. In: *Acute Diarrhoea in Childhood.* pp. 45–72. Edited by K. Elliot & J. Knight. Elsevier, Amsterdam.

TIMMIS, K. N., LEHRBACH, P. R., HARAYAMA, S., DON, R. H., MERMOD, N., BAS, S., LEPPIK, R., WEIGHTMAN, A. J., REINEKE, W. and KNACKMUSS, H. J. (1985) Analysis and manipulation of plasmid-encoded pathways for the catabolism of aromatic compounds by soil bacteria. In: *Plasmids in Bacteria.* pp. 719–739. Edited by D. R Helinski, S. N. Cohen, D. B. Clewell, D. A. Jackson, A. Hollaender & C. M. Wilson. Plenum, New York.

6 Plasmids and genetic engineering

Plasmids are used as *vectors* (or *vehicles*) to *clone* DNA molecules. Restriction endonucleases, ligase and other enzymes can be used to add pieces of foreign DNA, called *inserts,* to plasmid molecules *in vitro*. The recombinant plasmid can then be put back into a suitable host bacterium, usually *E. coli,* by transformation. The insert is added at an inessential site in the vector so that the recombinant plasmid can replicate in the bacterium.

The main steps in using plasmid vectors to clone DNA molecules are as follows:

a Isolation of circular plasmid DNA molecules and preparation of the DNA insert. The insert may be a piece of chromosomal DNA from an animal or plant, a cDNA (complementary DNA) molecule derived from mRNA, or a chemically synthesized sequence.
b Insertion of the foreign DNA into the plasmid. There are several ways of doing this, but they all involve breaking a circular plasmid at a specific point with a restriction endonuclease to convert it into a linear molecule. The cleaved plasmid is mixed with the DNA to be inserted (the ends of the insert are usually made homologous to the ends of the vector) and the two are sealed together by DNA ligase to form a circular molecule.
c Addition of the recombinant plasmids to host bacteria by transformation.
d Identification of transformants which have recombinant plasmids. Usually only a fraction of all the transformants contain the cloned fragments of interest.

Restriction endonucleases

There are two types of restriction endonuclease but only one of these, type II, is used extensively in genetic engineering. Typical type I enzymes are *Eco*B and *Eco*K. These are specified by chromosomal genes of *E. coli* strain B and *E. coli* K-12, respectively. They recognize a particular sequence of nucleotides and cut the DNA molecule. However, type I enzymes do not always cut at the same sequences even though they have a specific recognition sequence. The chromosomes of *E. coli* B and *E. coli* K-12 are protected from these endonucleases because the same enzymes *modify* the sequence they recognize by methylating particular nucleotides within it. However, an invading virus chromosome is much more likely to be cut than methylated by the enzymes. It is often assumed that restriction-modification systems have evolved to protect cells and the replicons they contain from viruses. The temperate bacteriophage P1 also specifies a type I restriction-modification system.

Type II restriction endonculeases differ from type I enzymes in that they not only recognize a particular sequence but also cut DNA at a particular site, usually within the recognition sequence. The type I enzymes found in *E. coli* strains B and

K-12 require Mg^{2+}, ATP and S-adenosyl methionine for endonuclease activity, whereas type II enzymes need only Mg^{2+} as a cofactor. Unlike type I enzymes, type II enzymes have only one type of protein subunit and this is not also involved in methylating DNA.

More than 470 different type II enzymes have been isolated so far from a wide range of bacterial species. They are given names which indicate their origin. Thus, *Hind*III is one of at least three restriction endonucleases produced by *Haemophilus influenzae* serotype d. It is not known whether all type II enzymes form part of restriction-modification systems, but the bacterial chromosomes must presumably be protected in some way from the effects of the restriction enzymes. Type II enzymes can be coded for by plasmids. One of the most commonly used enzymes is *Eco*R1 which is specified by a nonconjugative R plasmid related to ColE1 (not the conjugative R1 plasmid which is described in Chapter 4). Type II enzymes differ from each other in the types of nucleotide sequence they recognize and in the type of break they make in DNA.

The sequences recognized by type II enzymes can be classified into three groups as shown in Table 8. The sites recognized by most enzymes have a two-fold axis of symmetry and are said to be *palindromic*. The palindrome can be a tetra-, penta- or hexanucleotide sequence. Cleavage of DNA can leave either *blunt-ends* (as in the case of *Alu*I) or *sticky* (or *cohesive*) *ends* in which a few unpaired nucleotides, terminating in either a 3′hydroxyl (as in the case of P*st*I) or a 5′-monophosphate group (as in the case of *Eco*R1), project from the ends. Most of the restriction

Table 8 Recognition specificities of type II restriction endonucleases

palindromic sites					
tetranucleotides		pentanucleotides		hexanucleotides	
AG↓CT	(*Alu*I)	G↓ANTC	(*Hin*fI)	G↓AATTC	(*Eco*RI)
GG↑CC	(*Hae*III)	↓CC(A/T)GG	(*Eco*RII)	A↓AGCTT	(*Hind*III)
				CTGCA↓G	(*Pst*I)
				GTT↓AAC	(*Hpa*I)
asymmetric sites					
tetranucleotides		pentanucleotides		relaxed sites	
5′-CCTC(N)	5–10↓ (*Mnl*I)	5′-GACGC(N)	5↓ (*Hga*I)	PuGCGCPy↓	*Hae*II)
3′-GGAG(N)		3′-CTGCG(N)	10↑	GTPy↓PuAC	(*Hind*II)

Sequences are written from 5′ → 3′ and only one strand is shown. Cleavage sites are indicated by arrows. Thus the *Eco*R1 site, G↓AATTC, is an abbreviation for 5′G↓A-A-T-T-C-3′.
3′C-T-T-A-A↑G-5′.

From Zabeau, M. & Roberts, R.J. (1979). In: *Molecular Genetics Part III*. pp1–63. Edited by J.H. Taylor. Academic Press, New York.

endonucleases used in genetic engineering recognize tetra- or hexanucleotide palindromic sequences.

A few type II enzymes recognize asymmetric sites. These do not cleave DNA at positions within the recognition sequence but at a distance of five to ten base pairs to one side of it. The third class of enzymes have *relaxed* recognition sites and do not discriminate between different purines or pyrimidines at certain positions in their recognition sites.

A total of 36 different sequences are recognized by one or more type II enzymes. Enzymes from different bacteria which recognize the same sequence, are called *isoschizomers.* Although they recognize the same sequence, isoschizomers do not necessarily cut DNA at the same positions within the sequence.

If a DNA molecule has an (A + T):(G + C) ratio of 1, a particular hexanucleotide sequence should occur on average every 4096 (that is, 4^6) base pairs. The fragments produced by an enzyme recognizing a tetranucleotide sequence should have an average length of 256 base pairs. A typical prokaryotic gene is about 1 000 nucleotides long. Some DNA molecules have far fewer endonuclease-sensitive sites than would be expected. RP4 is a wide host-range plasmid comprising about 57 000 base pairs, but it has only single sites sensitive to *Eco*R1, *Hin*dIII, *Bam*HI and *Bgl*II, enzymes which recognize hexanucleotide sequences. All these sites are in, or close to, genes coding for antibiotic-resistance. It has been suggested that plasmids having a wide host-range are often exposed to restriction enzymes so there is a strong selection pressure for loss of sensitive sites.

Because of the specificity of type II enzymes, a plasmid or other DNA molecule is always cut at the same position by a given enzyme. It is often useful to have a map of the plasmid showing the positions of these sites. Maps of endonuclease-sensitive sites within F and several R plasmids have been determined by analysing the molecular weights of fragments generated by restriction endonucleases. Molecular weights are determined by electrophoresis through agarose or acrylamide gels; the positions of the fragments are compared with those of molecular weight standards. The gels are stained with ethidium bromide and illuminated with ultraviolet light to show the positions of the DNA fragments. The relative positions of sensitive sites within a molecule can be obtained by sequential digestion of fragments with different enzymes and by partial digestion of molecules so that some sensitive sites remain uncut. Comparison of fragments generated by restriction endonucleases can indicate similarities between plasmids. For example, TOL plasmids found in bacteria isolated from soil in Japan and in Wales gave identical fragments when digested with *Eco*R1.

Plasmid vectors

The most commonly used plasmid vectors are pBR322 and other derivatives of pMB1, a plasmid closely related to ColE1-K30. The host bacterium of these plasmids is *E. coli,* usually a derivative of *E. coli* K-12. A *disabled* strain of *E. coli,* which has mutations preventing it from growing in anything except specially prepared growth media, may be used if the recombinant plasmid is potentially

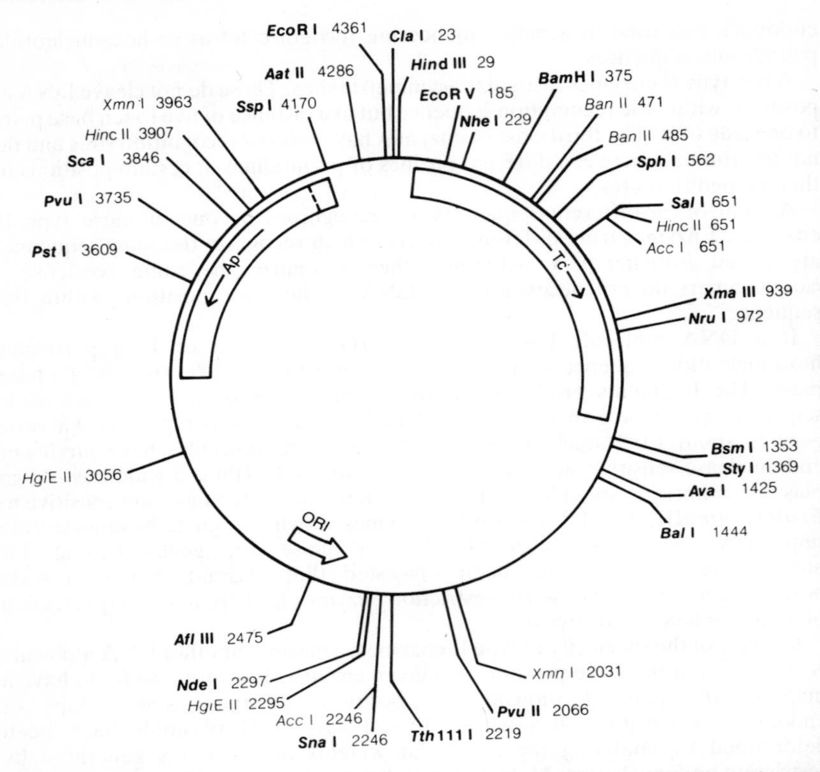

Fig. 21 pBR322 was made by combining DNA from Tn*3* (the ampicillin resistance gene), pSCl0l (the tetracycline resistance gene) and pMBl (the origin of replication). pMBl is a close relative of ColEl. The first T in the unique *Eco*RI-sensitive site . . . GA*T*TC . . . is designated as nucleotide number 1. pBR322 comprises 4363 base pairs. Enzymes which cut the molecule only once are shown in bold type; the other enzymes cut the molecule twice. (Figure from New England Biolabs.)

dangerous. Plasmid vectors have also been developed for use in *Bacillus*, *Streptomyces* and *Pseudomonas*.

The ColE1-type replicon has several advantages as the basis for a vector. ColE1 is a multicopy plasmid which continues to replicate—in fact it replicates at a faster rate—when chromosomal replication is inhibited by chloramphenicol (p.5). That is, ColE1 is *amplified* in chloramphenicol-treated cultures and can be easily isolated from such cultures. The essential replication region of ColE1-type plasmids is a relatively short sequence; small vectors are more useful because they are transformed at higher frequencies than large plasmids.

pBR322 (Fig. 21) consists of the essential replication region of pMB1 to which two drug-resistance genes, coding for ampicillin-resistance and tetracycline-resistance have been added by ligating fragments of R plasmids. The ampicillin-resistance gene is part of the transposon Tn*3* (from plasmid R1) which codes for a

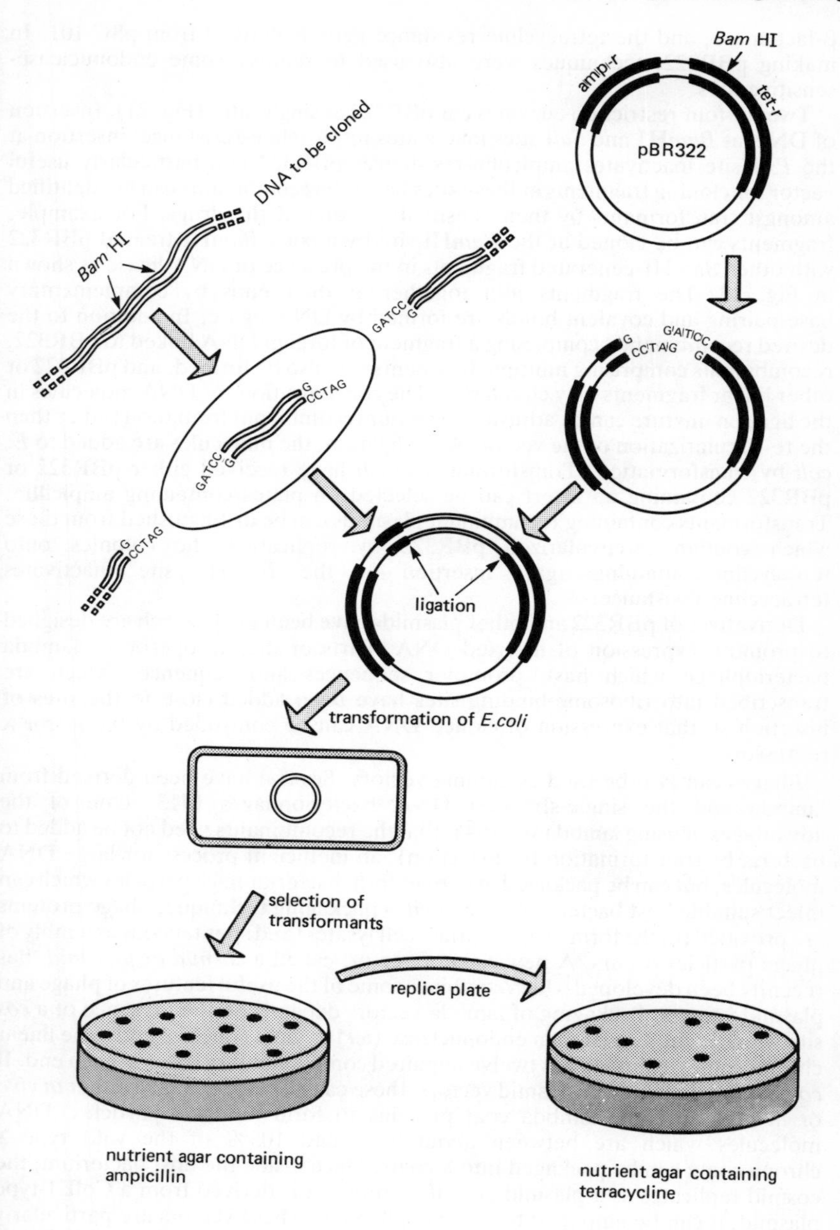

Fig. 22 Using pBR322 to clone DNA

β-lactamase, and the tetracycline resistance gene is derived from pSC 101. In making pBR322, techniques were also used to remove some endonuclease-sensitive sites.

Twenty-four restriction enzymes cut pBR322 at single sites (Fig. 21). Insertion of DNA at *Bam*H1 and *Sal*I sites inactivates tetracycline-resistance; insertion at the *Pst*I site inactivates ampicillin-resistance. pBR322 is a particularly useful vector for cloning fragments at these sites because recombinants can be identified amongst transformants by their sensitivity to one of the drugs. For example, fragments can be cloned at the *Bam*HI site by mixing *Bam*HI-treated pBR322 with other *Bam*HI-generated fragments in the presence of DNA ligase as shown in Fig. 22. The fragments join together at their ends by complementary base-pairing and covalent bonds are formed by DNA ligase. In addition to the desired recombinants, comprising a fragment of foreign DNA linked to pBR322, recombinants comprising multiple fragments can also be formed, and pBR322 or other linear fragments may circularize. The concentrations of DNA molecules in the ligation mixture can be adjusted to favour recombinant formation rather than the re-circularization of the vector. After ligation, the molecules are added to *E. coli* by transformation. Transformants which have received either pBR322 or pBR322 containing an insert can be selected on plates containing ampicillin. Transformants containing recombinant plasmids can be distinguished from those which contain re-circularized pBR322 by replicating the colonies onto tetracycline-containing agar; insertion at the *Bam*H1 site inactivates tetracycline-resistance.

Derivatives of pBR322 and other plasmids have been made which are designed to promote expression of inserted DNA. Parts of the *lac* operon or lambda bacteriophage which have promoter sequences and sequences which are transcribed into ribosome-binding sites have been added close to the sites of insertion so that expression of cloned DNA can be controlled by the *lac* or λ repressors.

Phages can also be used as cloning vectors. Several have been derived from lambda and the single-stranded DNA bacteriophages M13. One of the advantages of using lambda vectors is that the recombinants need not be added to bacteria by transformation (transfection), an inefficient process for large DNA molecules, but can be packaged *in vitro* to form bacteriophage particles which can infect suitable host bacteria. In the *in vitro* packaging technique, phage proteins are provided (in the form of appropriate cell lysates) and spontaneous assembly of phage particles occurs. A new class of vector, called a *cosmid* or *phasmid,* has recently been developed which combines some of the useful features of phage and plasmid vectors. Packaging of lambda vectors depends on the presence of a *cos* site which is cut by a lambda endonuclease (*ter*) to yield cohesive ends; the linear chromosome of lambda has twelve unpaired complementary bases at each end. If *cos* sites are included in plasmid vectors, these can also be packaged either *in vivo* or *in vitro* into the lambda coat proteins to form infective particles. DNA molecules which are between about 75% and 105% of the wild type λ chromosome can be packaged into λ coats. Once inside the host bacterium, the cosmid replicates as a plasmid and, if its replicon is derived from a ColE1-type plasmid, it can be amplified by chloramphenicol. These vectors are particularly useful for cloning long DNA molecules.

Methods used to link inserts to plasmid vectors

Sticky ends generated by restriction endonucleases This is perhaps the simplest method for cloning DNA (Fig 22). Both the insert and the vector are cut with the same restriction endonuclease which cleaves a palindromic sequence to leave sticky ends. These ends bind the two DNA molecules together and covalent bonds are formed by DNA ligase. The DNA ligase used is that specified by bacteriophage T4.

Different procedures are often necessary. Many inserts are not generated by restriction endonucleases but are complementary DNA (cDNA) molecules derived from RNA. The ends of such molecules must be tailored to make them suitable for joining to plasmid vectors. It may also be necessary to change the sequence at the ends of vectors or inserts to get efficient expression of the cloned DNA.

Blunt-end ligation and use of linkers DNA ligase can join DNA molecules which do not have projections of single-stranded DNA to hold them together, although ligation of blunt ends is much less efficient than ligation of sticky ends. The reaction can be used to join fragments made by restriction endonucleases, such as *Alu*I and *Hpa*I, which cut DNA to leave blunt ends (Table 8). Alternatively, sticky ends can be converted to blunt ends, either by using *E. coli* DNA polymerase I to add nucleotides complementary to a 5′ extension or by removing an extension with a suitable nuclease.

Blunt-end ligation can also be used to attach synthetic sequences, *linkers,* to the end of molecules. Subsequent digestion of the linker with the appropriate restriction endonuclease yields sticky ends which can be joined to the complementary sticky ends of another molecule. For example, the linker
CCAAGCTTGG
GCTTCGAACC, which contains a *Hin*dIII-sensitive sequence, can be added by blunt-end ligation to both ends of a DNA molecule. Subsequent digestion with *Hin*dIII leaves cohesive ends which can be joined to a *Hin*dIII-cleaved vector. The advantage of having a site for *Hin*dIII or a similar enzyme at each junction of insert and vector is that the cloned fragment can be re-isolated from the recombinant plasmid by digestion with the appropriate enzyme.

Extensions formed by terminal transferase Terminal deoxynucleotidyltransferase adds deoxynucleotide residues to the 3′ ends of DNA molecules. Complementary single-stranded extensions of vector and insert can be made by adding a tail of several deoxyguanosine residues(oligo(dG)) to the 3′ ends of the vector and oligo(dC) to the 3′ ends of the insert. When mixed together, the molecules become linked by complementary base-pairing. A disadvantage of this technique is that it is usually impossible to reisolate the insert from the plasmid recombinant by a simple procedure. However, if the insert is made by the oligo (dC)-oligo(dG) tailing procedure as a *Pst*I site (as in the β-lactamase gene of pBR322, for example), *Pst*I sites (see Table 8) are reformed on either side of the insert. An advantage of the tailing method is that the only circular molecules which are formed are recombinants. Recircularization of the plasmid vector cannot occur. (Recircularization of the vector can also be prevented, if necessary, in the ‘sticky ends' method described earlier by treatment with phosphatase to remove the 5′-phosphate groups from the vector prior to ligation.)

Transformation

When *E. coli* is treated with $CaCl_2$ it becomes competent and can take up DNA molecules by transformation. The usual procedure for adding plasmid DNA to *E. coli* by transformation is to obtain cells from a broth culture by centrifugation and then to resuspend them in 0.1M $CaCl_2$ at 0°C after a wash in $CaCl_2$. DNA is added and binds to the cells. After 5 min at 0°C, the suspension is kept at 37°C for 5 min during which time DNA is taken into the cells. The cells are then transferred to broth at 37°C to allow expression of transformed DNA and are spread onto agar containing appropriate drugs to select for transformants. The mechanism of $CaCl_2$-induced competence is not well understood.

Circular DNA molecules are transformed much more efficiently than linear molecules, and small circles are transformed more efficiently than large circles. Transformation of a population of different DNA molecules therefore does not yield a representative sample of the sizes of circular molecules present. The transformation frequency of pBR322 is about 10^6 transformants per μg of supercoiled DNA.

Identification of clones

The various enzymes and linkers which are used to clone DNA into plasmid vectors are available from commercial sources and it is often a relatively simple matter to obtain recombinant plasmids by *in vitro* techniques. However, it is sometimes much more difficult to identify the particular clones which have the gene or genes of interest. Much depends on the purity of the DNA to be cloned and on whether the insert is likely to be expressed in bacterial cells.

Methods dependent on expression of inserts Several genes from lower eukaryotes are expressed in *E. coli.* They are usually expressed inefficiently, however, presumably because eukaryotic signals for initiation of transcription and/or whether are not recognized efficiently in prokaryotic cells.

Recombinant plasmids containing certain genes from *Saccharomyces cerevisiae* and *Neurospora crassa* can be detected by complementation of equivalent genes in *E. coli.* For example, plasmids with inserts of chromosomal DNA from these two fungi can be used to transform an *E. coli* mutant which is defective in the production of an enzyme for, say, the synthesis of an amino acid. It is often found that recombinant plasmids with DNA from lower eukaryotes can complement the mutant *E. coli* so that the bacteria can be grown on media lacking the amino acid.

Many genes in higher eukaryotes have *introns* or *intervening sequences.* These are sequences which occur in the middle of a gene but do not code for amino acids which appear in the protein specified by the gene. Introns are removed by *splicing* from RNA transcripts of the gene. To obtain expression of eukaryotic genes in bacteria usually implies that the insert is a cDNA molecule derived from a mRNA from which the transcribed introns have been removed; *E. coli* does not produce a splicase to remove transcribed introns.

Recombinant plasmids containing the mouse gene for dihydrofolate reductase were detected simply by plating transformants onto media containing trimethoprim. The mammalian enzyme is more resistant than the bacterial enzyme to this drug. Reverse transcriptase from avian myeloblastosis virus was used to synthesize the DNA complementary to an RNA preparation which was enriched for dihydrofolate reductase mRNA. The RNA strand was removed from The RNA/DNA hybrid molecules by hydrolysis with alkali and the complementary DNA strand was synthesized in a reaction catalysed by DNA polymerase I (from *E. coli*) to produce double-stranded cDNA. Oligo (dC) extensions were added to the 3′ ends of the cDNA fragments, which were than annealed to pBR322 which had been cut with *Pst*I and had oligo (dG) extensions on its 3′ ends. Transformants which grew on plates containing trimethoprim were found to have recombinant plasmids which specified the mouse enzyme.

Detecting cloned mammalian genes is usually more difficult than this. Immunological methods are useful if antibodies against the protein which is made by the recombinant plasmid are available. Several eukaryotic proteins have been made in bacteria by fusing a eukaryotic gene to a prokaryotic gene; the eukaryotic proteins produced by these cells are fused to parts of bacterial proteins. Fused peptides which have been made include rat pre-growth hormone and rat proinsulin linked to part of the β-lactamase (penicillinase) peptide, and chicken ovalbumin linked to β-galactosidase. The eukaryotic parts of these fused peptides retain antigenic activity. Radio-immunoassay techniques have been developed for screening individual colonies of transformants for the production of eukaryotic proteins. Once a recombinant plasmid containing the gene of interest has been identified, further genetic manipulation can be used to produce a separate eukaryotic peptide.

Hybridization with RNA, DNA or synthetic oligonucleotides Transformants containing a recombinant plasmid can be identified by hybridization with RNA or DNA homologous to the cloned DNA. The colonies of transformants to be tested are grown on nitrocellulose filters placed on nutrient agar. A reference set of colonies is prepared by replica plating. Colonies on the filters are lysed to release their DNA which is denatured and becomes fixed to the part of the filter originally occupied by the colony. Radioactive RNA or DNA homologous to the DNA to be detected is then used as a *probe*. When denatured and added to the filter it binds to those recombinants which have homologous DNA. After washing the filters to remove non-specifically bound probe, the positions of the radioactive probe can be determined by autoradiography.

Synthetic oligonucleotides can be particularly useful for identifying specific DNA or RNA sequences, particularly when their use is coupled with new techniques for sequencing very small amounts of protein. The first step in the procedure is to obtain some sequence of the protein of interest, perhaps the N-terminal sequence or an internal sequence from a fragment obtained by proteolytic digestion. A part of the sequence is then chosen which has the least codon degeneracy. Only methionine and tryptophan are encoded by unique codons so there will inevitably be ambiguities in the deduced DNA sequence of most peptides. Oligonucleotides (usually comprising 14–23 nucleotides) are then synthesized corresponding to all or some of the coding sequences. The pool of radioactive oligonucleotide probes can then be used to screen colonies or cosmid plaques for homologous sequences.

Clones can also be identified by testing their effects on a cell-free system for translation of mRNA. If the cloned DNA hybridizes with mRNA in the system, production of a particular protein will be inhibited.

Summary

Small recombinant plasmids, such as pBR322, which have genes coding for drug resistance are widely used as vectors for cloning DNA molecules. DNA is usually inserted at a site which inactivates one of the resistance genes so that recombinant plasmids can be detected. Phages and cosmids, which are made by combining parts of plasmids and phage DNA, are also used as vectors.

Methods used to insert DNA into vectors

a Cleavage of both vector and insert with a type II restriction endonuclease which leaves complementary single-stranded sequences at the ends of both molecules. T4 DNA ligase is used to form covalent bonds between the molecules.

b Ligation of fragments produced by enzymes which leave blunt ends, or addition of synthetic oligonucleotide linkers to the fragment to be cloned. The linker can be digested with an enzyme which leaves cohesive ends so that the DNA can be inserted as in (a).

c Addition of complementary deoxyribonucleotide residues to the 3′ ends of the insert and vector. For example, a tail of oligo (dG) can be added to the vector and oligo (dC) to the insert.

Transformation Recombinant plasmids are added to a suitable host bacterium (*E. coli* if pBR322 is used) by transformation. *E. coli* is treated with $CaC1_2$ to enable it to take up plasmids.

Identification of clones

a Complementation of bacterial genes Some genes from lower eukaryotes can complement mutant bacterial genes; transformants containing these genes can be detected simply by spreading bacteria on plates containing suitable growth media. Many genes from higher eukaryotes have introns which are removed, by splicing, from mRNA. To obtain expression of these genes a complementary (cDNA) molecule is obtained from mRNA, or the gene is synthesized chemically if the amino acid sequence of the protein is known. A cloned cDNA copy of the mouse gene for dihydrofolate reductase was expressed in *E. coli* so that transformants could be selected on plates containing trimethoprim, to which the mouse enzyme is resistant.

b Immunological methods can be used to detect clones producing a eukaryotic protein. This technique can be successful even when the eukaryotic protein is fused to part of a prokaryotic peptide.

c Hybridization with RNA or DNA synthetic oligo nucleotide. Colonies of transformants growing on nitrocellulose filters are lysed to release their denatured DNA which binds to the filters. A radioactive probe of RNA, DNA or synthetic oligonucleotide is then added and binds to homologous DNA on the filter. The positions of the probe can be determined by autoradiography. Recombinant plasmids can also be identified by their inhibitory effects on a system for translation of mRNA.

References

OLD, R. W. and PRIMROSE, S. B. (1985). *Principles of genetic manipulation.* Blackwell Scientific Publications, Oxford.

Plasmid	Original host	Molecular weight	Copy no	Incompatibility group	Conjugation	Characteristics
F	*Escherichia coli* K-12	63×10^6	1–2	FI	+ (F pili)	Integrates to form Hfr strains and can form F′ plasmids. *Tra* operon naturally de-repressed.
R1	*Salmonella paratyphi*	58×10^6	1–2	FII	+ (F pili)	Dissociates into a conjugative resistance-transfer factor (RTF) and r-determinant (Ap. Cm, Fa, Km, Sm, Su). R1*drd*19 is de-repressed for conjugal transfer.
R100	*Shigella flexneri*	58×10^6	1–2	FII	+ (F pili)	Also called NR1 and R222. Tc (Tn*10* on RTF) and Cm, Fa, Hg, Sm, Su (on r-determinant). R-100.1 is de-repressed for conjugal transfer.
R6	*Escherichia coli*	66×10^6	1–2	FII	+ (F pili)	Tc (Tn*10* on RTF) and Cm. Fa, Hg, Sm, Su (on r-determinant). R6–5 is a tetracycline-sensitive derivative.
R6K	*Proteus rettgeri*	25×10^6	about 15	X	+	Ap, Sm. Bidirectional, asymmetric replication.
RP1, RP4, RK2 R68	*Pseudomonas aeruginosa* *Klebsiella aerogenes* *Pseudomonas aeruginosa*	38×10^6	1–2	PI	+	Ap (Cb), Km, Tc. Wide host-range. RP1, RP4 and RK2 isolated from same Birmingham (UK) hospital and probably identical. R68 is very similar. R68.45 transfers chromosomal genes more frequently.

Co1V,I-K94	*Escherichia coli*	85×10^6	1–2	FI	+ (F pili)	Colicins V and I. Complement-resistance. Naturally de-repressed *tra* operon.
Co1B-K98	*Escherichia coli*	70×10^6	1–2	FIII	+ (F pili)	Colicin B.
Co1Ib-P9	*Shigella sonnei*	65×10^6	1–2	Iα	+ (I pili)	Colicin I; prototype I sex pili.
Δ	*Salmonella typhimurium* Phage type 29	61×10^6	1–2	Iα	+ (I pili)	Mobilizes non-conjugative R plasmids found in same strain.
ColE1-K30	*E. coli*	4.3×10^6	about 15	?	—	Colicin E1.
BR322		2.9×10^6	about 40	?	—	Cloning vector. Ap, Tc. Made by combining parts of R1, pSC101 and pMBI.
pSC101	*Salmonella panama*	6×10^6	about 5	?	—	Tc
TOL (pWWO)	*Pseudomonas putida*	78×10^6	?	?	+	Degradative plasmid (toluene catabolism)
Ti plasmids	*Agrobacterium tumefaciens*	$90–160 \times 10^6$	?	?	+	Crown Gall disease in dicotyledonous plants

A double-stranded DNA molecule comprising 1500 base pairs has a molecular weight of about 1×10^6 and a length of about 2.1 μm. 1000 bases codes for a polypeptide of 333 amino acid residues (molecular weight about 30 000).

Key to symbols for resistances: Ap, ampicillin; Cb, carbenicillin; Cm, chloramphenicol;Fa, fusidic acid; Hg, mercuric ions; Km, kanamycin; Sm, streptomycin; Su, sulphonamide; Tc, tetracycline.

Index